Curriculum-Based Activities and Resources for Preservice Math Teachers

Curriculum-Based Activities and Resources for Preservice Math Teachers

Edited by

Gwendolyn M. Lloyd
The Pennsylvania State University

Vanessa R. Pitts Bannister
University of South Florida Polytechnic

NATIONAL COUNCIL OF TEACHERS OF MATHEMATICS

www.nctm.org/more4u
Access code: CBA13993

Copyright © 2011 by
The National Council of Teachers of Mathematics, Inc.
1906 Association Drive, Reston, VA 20191-1502
(703) 620-9840; (800) 235-7566; www.nctm.org
All rights reserved

Library of Congress Cataloging-in-Publication Data

Curriculum-based activities and resources for preservice math teachers / edited by Gwendolyn M. Lloyd, Vanessa R. Pitts Bannister.
  p. cm.
  ISBN 978-0-87353-682-0
  1. Mathematics teachers—Training of. 2. Mathematics—Study and teaching—Activity programs. 3. Mathematics—Study and teaching—Standards. I. Lloyd, Gwendolyn M. II. Pitts Bannister, Vanessa R.
  QA10.5.L436 2010
  510.71—dc22
                    2010052728

The National Council of Teachers of Mathematics is a public voice of mathematics education, supporting teachers to ensure equitable mathematics learning of the highest quality for all students through vision, leadership, professional development, and research.

For permission to photocopy or use material electronically from *Curriculum-Based Activities and Resources for Preservice Math Teachers,* please access www.copyright.com or contact the Copyright Clearance Center, Inc. (CCC), 222 Rosewood Drive, Danvers, MA 01923, 978-750-8400. CCC is a not-for-profit organization that provides licenses and registration for a variety of users. Permission does not automatically extend to any items identified as reprinted by permission of other publishers and copyright holders. Such items must be excluded unless separate permissions are obtained. It will be the responsibility of the user to identify such materials and obtain the permissions.

The publications of the National Council of Teachers of Mathematics present a variety of viewpoints. The views expressed or implied in this publication, unless otherwise noted, should not be interpreted as official positions of the Council.

Printed in the United States of America

Contents

Preface vii
      Gwendolyn M. Lloyd
      The Pennsylvania State University, University Park, Pennsylvania
      Vanessa R. Pitts Bannister
      University of South Florida Polytechnic, Lakeland, Florida

I. Introduction 1

1. Textbooks and Curriculum Materials in Mathematics Teacher Education: Preservice Teachers' Learning about Important Principles for School Mathematics 3
      Gwendolyn M. Lloyd
      The Pennsylvania State University, University Park, Pennsylvania

II. Principles for School Mathematics 11

2. The Equity Principle: Developing Preservice Secondary School Mathematics Teachers' Conceptions of Equity—an Application of Standards-Based Curriculum Materials 13
      Vanessa R. Pitts Bannister
      University of South Florida Polytechnic, Lakeland, Florida
      Gina J. Mariano
      Troy University, Troy, Alabama
            Activity 2.1 Equity and Curriculum Materials: High Expectations 19
            Activity 2.2 Equity and Curriculum Materials: Differentiation 20

3. The Curriculum Principle: Looking through Lenses—a Tripartite Examination of Textbooks 23
      Beth A. Herbel-Eisenmann
      Michigan State University, East Lansing, Michigan
            Activity 3.1 Example Project from a Methods Course Focused on Algebra 32
            Activity 3.2 Analyzing Tasks' Cognitive Demand 33

4. **The Teaching Principle: Learning to Address the Teaching Principle with Standards-Based Curriculum Materials** 35
   Corey Drake
   Iowa State University, Ames, Iowa
   Tonia J. Land
   Iowa State University, Ames, Iowa
   - Activity 4.1   In-Class Teaching 46
   - Activity 4.2   Analyzing Important Fractions Content in Two Curriculum Series 48
   - Activity 4.3   Analyzing Curriculum Materials with a Variety of Criteria 50

5. **The Learning Principle: Supporting the Development of Mathematical Proficiency** 55
   Fran Arbaugh
   The Pennsylvania State University, University Park, Pennsylvania
   - Activity 5.1   Task Analysis: Supporting Mathematical Proficiency 64
   - Activity 5.2   Anticipating Students' Responses 66

6. **The Assessment Principle: Broadening Preservice Teachers' Views of Assessment through Engagement with Curriculum Materials** 69
   Gwendolyn M. Lloyd
   The Pennsylvania State University, University Park, Pennsylvania
   - Activity 6.1   Looking for Evidence of Students' Understanding 77
   - Activity 6.2   Examining Assessment Resources in Curriculum Materials 78

7. **The Technology Principle: Using Standards-Based Curriculum Materials to Learn to Teach Mathematics with Technology** 81
   Susan M. Hagen
   Virginia Polytechnic Institute and State University, Blacksburg, Virginia
   - Activity 7.1   Mathematics Teaching Project 90
   - Activity 7.2   Are All Constructions Created Equal? 93

III. From the University to the Classroom 99

    8. **Preservice Teachers' Learning in Grades K–12 Classrooms: Engaging with Standards-Based Mathematics Curriculum Materials in Field Experiences** 101

        Andrea V. McCloskey
        The Pennsylvania State University, University Park, Pennsylvania
        Elizabeth Winarski
        Bloomington Project School, Bloomington, Indiana

IV. Resources 109

    9. **Teacher Educators' Access to Curriculum Resources: History and Development of Standards-Based Programs, Sample Materials, and Technological Tools** 111

        Gwendolyn M. Lloyd
        The Pennsylvania State University, University Park, Pennsylvania

Author Biographies 117

Preface

Gwendolyn M. Lloyd
Vanessa R. Pitts Bannister

In response to the publication of the Curriculum Standards (National Council of Teachers of Mathematics [NCTM] 1989), more than a dozen mathematics curriculum programs were developed during the 1990s, and millions of students have used these programs in grades K–12 classrooms over the past decade (Senk and Thompson 2003). A sizable collection of professional books related to these Standards-based curriculum programs has been published in recent years. These books address, for example, the selection of curriculum materials (Goldsmith, Mark, and Kanstrov 2000), the curriculum development process (Hirsch 2007), and the broad trends and issues that characterize this significant period of curriculum change in mathematics education (NCTM 2010). Two research volumes report about evaluations of Standards-based curriculum materials, with particular focus on students' learning (National Research Council 2004; Senk and Thompson 2003), and more recently, Remillard, Herbel-Eisenmann, and Lloyd (2009) compiled research about teachers' use of mathematics curriculum materials. Responding to our field's growing interest in relationships between teachers and textbooks, particularly ways that teachers learn through engagement with textbooks and curriculum materials, this book focuses on the role of Standards-based materials in preparing mathematics teachers.

This book represents the first compilation of strategies for using Standards-based curricula in mathematics teacher education. Part I introduces the notion of using Standards-based curriculum materials in mathematics teacher education. In part II, Principles for School Mathematics, teacher educators share ways of using Standards-based curriculum materials to challenge preservice teachers' views and understandings of NCTM's (2000) six Principles for school mathematics. These chapters show how preservice teachers can enhance their views of *equity, curriculum, teaching, learning, assessment*, and *technology* through different forms of engagement with textbooks and curriculum materials. These chapters also present strategies and specific activities, as well as teacher educators' accounts of preservice teachers' experiences learning with curriculum materials. Part III, From the University to the Classroom, extends the consideration of mathematics teacher education from university coursework to the mathematics classroom. Here, a mentor teacher and teacher educator examine the role of Standards-based curriculum materials in interns' learning during field experience components of teacher education programs. Part IV, Resources, offers an overview of the range of resources, available through the Internet, that can support teacher educators as they use Standards-based curriculum materials with preservice mathematics teachers.

We wrote this book primarily as a resource for mathematics teacher educators, particularly those who teach preservice teachers in university methods and mathematics courses. Although the activities we suggest herein were developed for use with preservice teachers, many of the activities can be adjusted for use with more experienced teachers in professional development settings as well. Adaptations of these activities may be effective for use in workshops with teachers unfamiliar with a newly adopted curriculum program or with parents and other community members interested in learning about curriculum programs. Although this book's primary audience is those involved with preparing mathematics teachers, the book may also appeal to researchers interested in teachers' learning. Each chapter contains many research questions, stated sometimes implicitly and other times explicitly, about preservice teachers' developing relationships with curriculum materials and understandings of mathematics teaching and learning. We hope that the insights and experiences that this book shares will contribute to both the practices of and future research about mathematics teacher education.

The National Science Foundation (grant number 0536678) supported this book's development. The opinions and recommendations expressed herein are those of the chapter authors and do not necessarily express the position or support of the foundation.

This book represents a collaboration among university- and school-based mathematics teacher educators who work in different teacher education programs but share the view that Standards-based curriculum materials hold tremendous potential for the learning of preservice elementary and secondary school mathematics teachers' learning. We thank the chapters' authors for their insightful contributions and willingness to share their teacher education practices and experiences with others, as well as for responding thoughtfully to our requests for revisions to increase the book's quality and coherence. We are also grateful to our former graduate students, Stephanie Behm Cross, Karl Kosko, and Gina Mariano, who contributed to this book's preparation while they and we were at Virginia Tech. Finally, we thank our colleagues, friends, and families for their support and encouragement during this project.

References

Goldsmith, Lynn T., June Mark, and Ilene Kanstrov. *Choosing a Standards-Based Mathematics Curriculum*. Portsmouth, N.H.: Heinemann, 2000.

Hirsch, Christian R. *Perspectives on the Design and Development of School Mathematics Curricula*. Reston, Va.: National Council of Teachers of Mathematics, 2007.

National Council of Teachers of Mathematics (NCTM). *Curriculum and Evaluation Standards for School Mathematics*. Reston, Va.: NCTM, 1989.

———. *Principles and Standards for School Mathematics*. Reston, Va.: NCTM, 2000.

———. *Mathematics Curriculum: Issues, Trends, and Future Directions*. 2010 Yearbook of the National Council of Teachers of Mathematics (NCTM). Reston, Va.: NCTM, 2010.

National Research Council. *On Evaluating Curricular Effectiveness: Judging the Quality of K–12 Mathematics Evaluations*. Washington, D.C.: National Academies Press, 2004.

Remillard, Janine T., Beth A. Herbel-Eisenmann, and Gwendolyn M. Lloyd. *Mathematics Teachers at Work: Connecting Curriculum Materials and Classroom Instruction*. New York: Routledge, 2009.

Senk, Sharon L., and Denisse R. Thompson. *Standards-Based School Mathematics Curricula: What Are They? What Do Students Learn?* Mahwah, N.J.: Lawrence Erlbaum Associates, 2003.

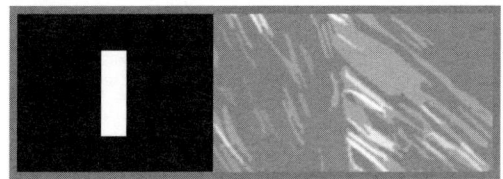

Introduction

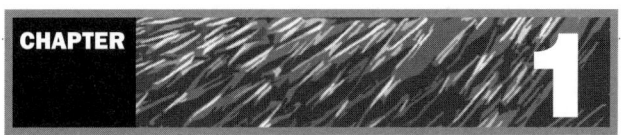

Textbooks and Curriculum Materials in Mathematics Teacher Education: Preservice Teachers' Learning about Important Principles for School Mathematics

Gwendolyn M. Lloyd

To support the vision of mathematics teaching and learning presented in the National Council of Teachers of Mathematics (NCTM) *Curriculum and Evaluation Standards for School Mathematics* (NCTM 1989), more than a dozen sets of grades K–12 mathematics curriculum materials were developed during the 1990s (National Research Council 2004; Senk and Thompson 2003). These Standards-based curriculum materials have affected millions of students' classroom experiences in U.S. schools and introduced new challenges and opportunities to teachers' work (Remillard, Herbel-Eisenmann, and Lloyd 2009). Implementation of curriculum materials varies considerably as teachers make different interpretations and develop personal ways of using the materials in diverse contexts. Engagement with curriculum materials also appears to affect teachers' mathematical and pedagogical understandings profoundly (Lloyd 1999; Remillard 2000; Remillard and Bryans 2004; Sherin and Drake 2009). The notion that curriculum materials can support and promote teachers' learning has contributed to a growing interest among teacher educators in using grades K–12 materials as the basis for early educative experiences for preservice teachers (PTs).

Engagement with textbooks and curriculum materials is a widely used professional development strategy for in-service teachers (Loucks-Horsley et al. 2010). Increasingly, mathematics teacher educators are using grades K–12 curriculum materials, particularly Standards-based curriculum materials, in preparing elementary and secondary school PTs. This book gives our community a collection of strategies and activities for using Standards-based curriculum materials in mathematics teacher education. The book also offers personal accounts from mathematics teacher educators about how and why they use Standards-based curriculum materials in their teacher education work, as well as reflections about how their PTs learned through engagement with curriculum materials.

Principles for School Mathematics

In *Principles and Standards for School Mathematics,* NCTM (2000) identified six Principles for school mathematics that communicate perspectives about a broad set of issues in mathematics education: equity, curriculum, teaching, learning, assessment, and technology. Figure 1.1 briefly describes each of these Principles. Taken together, they offer a useful framework for thinking about important themes and concerns in school mathematics.

Equity	Excellence in mathematics education requires equity—high expectations and strong support for all students.
Curriculum	A curriculum is more than a collection of activities: it must be coherent, focused on important mathematics, and well articulated across the grades.
Teaching	Effective mathematics teaching requires understanding what students know and need to learn and then challenging and supporting them to learn it well.
Learning	Students must learn mathematics with understanding, actively building new knowledge from experience and prior knowledge.
Assessment	Assessment should support the learning of important mathematics and furnish useful information to both teachers and students.
Technology	Technology is essential in teaching and learning mathematics; it influences the mathematics that is taught and enhances students' learning.

Fig. 1.1. Principles for school mathematics (NCTM 2000, p. 11)

In part II of this book, NCTM's six Principles for school mathematics serve as an organizing framework around which we share a variety of teacher education strategies that involve using Standards-based curriculum materials in courses for PTs. In chapter 2, Pitts Bannister and Mariano describe their efforts to develop PTs' understandings of the Equity Principle by using Standards-based curriculum materials in secondary school methods courses. In chapter 3, Herbel-Eisenmann shares three curriculum-based assignments that she developed to engage secondary school PTs in grappling with the issues raised in the Curriculum Principle. The authors of chapter 4, Drake and Land, report about different types of tasks—in-class teaching, using materials to determine important content, and analyzing curriculum materials—that they have designed and used to engage elementary school PTs with essential elements of the Teaching Principle. In chapter 5, Arbaugh presents a framework that helps PTs examine curriculum materials for how the tasks support students' learning. She also shares two activities that she has used in secondary school methods courses to highlight the Learning Principle. In chapter 6, about the Assessment Principle, Lloyd shares four activities that she has used in mathematics courses for elementary and secondary school PTs. These activities are intended to expand PTs' views of assessment and its role in mathematics teaching and learning. In chapter 7, Hagen offers examples of how she has used Standards-based curriculum materials in an undergraduate mathematics course to engage secondary school PTs in learning about the Technology Principle.

Part III, From the University to the Classroom, extends the consideration of mathematics teacher education from university coursework to grades K–12 mathematics classrooms. In chapter 8, McCloskey and Winarski examine the role of Standards-based curriculum materials in interns' learning during field experience components of teacher education programs. Finally, in part IV, Resources, chapter 9 gives an overview of the range of resources, available through the Internet, that can support teacher educators as they use Standards-based curriculum materials with PTs.

Contributions to Mathematics Teacher Education

Teacher Education Activities

Teacher education courses and field experiences can use Standards-based curriculum materials in many different ways and for many different purposes. Part II offers a diverse collection of strategies for using curriculum materials in teacher education courses. Some authors share strategies that they have created and used in mathematics courses, whereas others write about activities that PTs have completed in methods courses. Some chapters offer examples of elementary school PTs' learning, whereas others focus on that of secondary school PTs. These differences contribute to the richness and potential usefulness of this book's collection of strategies. As they explore the strategies shared in part II, readers will recognize that they can adapt these diverse activities for use with different teachers, in different settings, and for different purposes.

Common to all the chapters is an interest in using Standards-based curriculum materials to support PTs as they learn to teach mathematics in the spirit of the six Principles for school mathematics. The rich mathematical content and innovative instructional design of Standards-based curriculum materials give rise to potent opportunities for mathematics teacher educators to address a wide range of teaching and learning issues. In fact, most of part II's strategies can help accomplish multiple purposes related to not just one but several of the interconnected Principles (NCTM 2000). Part II's chapters contribute to the knowledge base in mathematics teacher education by sharing not only specific activities, which other educators can use or adapt, but also the thinking behind these activities' design and reports of experiences from PTs who have engaged in them.

The authors of chapters 2–7 have purposefully designed teacher education activities that not only capitalize on the representations of mathematics and pedagogy in Standards-based curriculum materials but also extend the use of curriculum materials beyond how grades K–12 students typically use them. Certainly teacher educators often ask PTs to complete selected problems or tasks *as learners* before the PTs reflect on and examine them *as teachers*. Teacher educators, however, have selected those tasks and problems carefully—from the wide range of curricular tasks originally written for grades K–12 students—so that the PTs will find them relevant and useful. Lloyd (2006, p. 17) comments as follows:

> Mathematics problems that provide learning opportunities for children are not necessarily productive or educative problems for consideration by teachers. The greatest potential for prospective teachers' learning about mathematics is through engagement with tasks that are mathematically challenging, emphasize conceptual understandings, address common misconceptions, and have potential to illustrate connections among concepts, representations, and real-world contexts. Tasks with greatest potential to illustrate or question important teaching and learning issues are those with multiple solution strategies, technology use, meaningful real-world contexts, problem-based learning, experimentation, and investigation.

In part II, we gain access to teacher educators' work as they strategically draw on and adapt the mathematical development and instructional design of Standards-based curriculum materials—and, in some instances, commercially developed textbooks—to design opportunities that support PTs' learning.

Activities in part II include analyzing Standards-based curriculum materials by using particular frameworks or issues as lenses, comparing Standards-based curriculum materials to commercially developed textbooks, and designing instruction (e.g., for peer teaching) on the basis of Standards-based curriculum materials. Such activities offer PTs opportunities to become familiar with representations of mathematical subject matter and pedagogical practices that are probably new to them, as well as to consider subject matter and pedagogy together, in relation to each other (Ball 2000). These activities can also contribute to developing PTs' awareness of the mathematical and instructional design of different curriculum materials. Through analysis of materials from a variety of curriculum programs and drawing on a range of frameworks and perspectives, PTs begin to recognize that curriculum developers make deliberate choices, reflecting different philosophies of teaching and learning, when they design instruction about a particular mathematical topic. Deep understanding of curricular features and design helps to position PTs to make reasoned decisions about selecting and adapting curriculum materials for their future classrooms.

In addition to offering curriculum-based activities for use by teacher educators in university courses for PTs (chapters 2–7) and identifying resources to support those activities (chapter 9), this book contributes insights into the role of curriculum materials in PTs' field experiences. Part III challenges university- and school-based teacher educators to consider how they might collaborate to offer a series of curriculum-based learning experiences, beginning in mathematics and methods courses and continuing into early field experiences and student teaching. The notion that curriculum materials may productively unify coursework with field experiences is an important contribution, worthy of further consideration in mathematics teacher education.

Research in Mathematics Teacher Education

Although this book does not aim to report research, the work it reports is situated within current research in mathematics teacher education. Research study findings and theoretical ideas have been the basis for, or have influenced, most of part II's teacher education activities. For example, activities in several chapters draw on a framework about tasks' cognitive demand (e.g., Smith and Stein 1998). Research's influence on part II's strategies and activities is perhaps not surprising: many chapter authors are both mathematics education researchers and practitioners in mathematics teacher education. Herbel-Eisenmann, for example, describes in chapter 3 how one way in which she invites PTs to examine curriculum materials grows out of her own research about how texts construct a "model reader." Drake and Land, in chapter 4, draw on their ongoing research about PTs' experiences with curriculum materials to offer evidence of PTs' learning.

The strategies that part II shares are not just closely tied to empirical findings and theoretical ideas: they also raise many issues and questions of potential interest and importance to researchers in mathematics teacher education. Whereas some chapters explicitly include questions or areas for future investigation, others imply questions through their strategies' and activities' purposes and results. The following are examples of questions that might interest readers as they read about strategies for using curriculum materials to develop PTs' understandings of the six Principles for school mathematics:

- What are PTs' conceptions of equity? How do teacher education activities involving Standards-based curriculum materials affect PTs' understandings of high expectations and differentiation? (Chapter 2, the Equity Principle)

- How do PTs' task classifications (e.g., according to levels of cognitive demand) change over time as PTs gain experience with different textbooks and curriculum materials and with different students and teachers? (Chapter 3, the Curriculum Principle)

- How does using curriculum materials to design and teach mathematics lessons to peers in methods courses influence PTs' ability and inclination to adapt curriculum materials? (Chapter 4, the Teaching Principle)

- How might exposure to several frameworks for classifying tasks influence PTs' initial efforts to choose tasks to support students' learning? (Chapter 5, the Learning Principle)
- How does using rubrics to analyze samples of students' work affect PTs' early efforts to design holistic rubrics? (Chapter 6, the Assessment Principle)
- How does learning or relearning mathematics with technology affect PTs' views about what mathematics should be taught and how it should be taught? (Chapter 7, the Technology Principle)

Inquiry into such questions could contribute to a small but growing body of research about PTs' learning with grades K–12 textbooks and curriculum materials during teacher education coursework (e.g., Frykholm 2005; Lloyd 2006; Lloyd and Behm 2005; Nicol and Crespo 2006; Tarr and Papick 2004).

In part III, we encounter the question of mathematics curriculum materials' role in PTs' field experiences, including student teaching. Several researchers have begun to examine how PTs use curriculum materials in their early efforts to design and enact mathematics instruction (Lloyd 2008; Nicol and Crespo 2006; van Zoest and Bohl 2002). This question is important, as McCloskey and Winarski suggest in chapter 8, because growing numbers of teacher education courses engage PTs in textbook analysis and adaptation, as well as in using Standards-based curriculum materials. What interactions do PTs have with the mathematics textbooks and curriculum materials, both Standards based and commercially developed, that they encounter in their early field experiences and student-teaching internships? How might use of Standards-based curriculum materials support PTs' efforts to design instruction that is consistent with the recommendations of the Common Core State Standards (CCSSI 2010)? How do curriculum-based experiences in teacher education courses prepare PTs to continue learning from using textbooks and curriculum materials and developing effective mathematics instruction in grades K–12 classrooms? Research about these questions could yield useful information about ways to design coursework and field experiences that support PTs' learning. Such research could also contribute to our broader understandings of how relationships among teachers, curriculum resources, and mathematics instruction develop over time.

References

Ball, Deborah L. "Bridging Practices: Intertwining Content and Pedagogy in Teaching and Learning to Teach." *Journal of Teacher Education* 51 (May–June 2000): 241–47.

Common Core State Standards Initiative (CCSSI). *Common Core State Standards for Mathematics.* Washington, D.C.: National Governors Association Center for Best Practices and the Council of Chief State School Officers, 2010. http://www.corestandards.org.

Frykholm, Jeffrey A. "Innovative Curricula: Catalysts for Reform in Mathematics Teacher Education." *Action in Teacher Education* 26 (Winter 2005): 20–36.

Lloyd, Gwendolyn M. "Two Teachers' Conceptions of a Reform-Oriented Curriculum: Implications for Mathematics Teacher Development." *Journal of Mathematics Teacher Education* 2 (October 1999): 227–52.

———. "Using K–12 Mathematics Curriculum Materials in Preservice Teacher Education: Rationale, Strategies, and Teachers' Experiences." In *The Work of Mathematics Teacher Educators: Continuing the Conversation,* AMTE Monograph 3, edited by Kathleen Lynch-Davis and Robin Ryder, pp. 11–28. San Diego: Association of Mathematics Teacher Educators, 2006.

———. "Curriculum Use While Learning to Teach: One Student Teacher's Appropriation of Mathematics Curriculum Materials." *Journal for Research in Mathematics Education* 39 (January 2008): 63–94.

Lloyd, Gwendolyn M., and Stephanie L. Behm. "Preservice Elementary Teachers' Analysis of Mathematics Instructional Materials." *Action in Teacher Education* 26 (Winter 2005): 48–62.

Loucks-Horsley, Susan, Katherine E. Stiles, Susan Mundry, Nancy Love, and Peter W. Hewson. *Designing Professional Development for Teachers of Science and Mathematics.* 3rd ed. Thousand Oaks, Calif.: Corwin Press, 2010.

National Council of Teachers of Mathematics (NCTM). *Curriculum and Evaluation Standards for School Mathematics.* Reston, Va.: NCTM, 1989.

———. *Principles and Standards for School Mathematics.* Reston, Va.: NCTM, 2000.

National Research Council. *On Evaluating Curricular Effectiveness: Judging the Quality of K–12 Mathematics Evaluations.* Washington, D.C.: National Academies Press, 2004.

Nicol, Cynthia C., and Sandra M. Crespo. "Learning to Teach with Mathematics Textbooks: How Preservice Teachers Interpret and Use Curriculum Materials." *Educational Studies in Mathematics* 62 (July 2006): 331–55.

Remillard, Janine T. "Can Curriculum Materials Support Teachers' Learning? Two Fourth-Grade Teachers' Use of a New Mathematics Text." *Elementary School Journal* 100 (March 2000): 331–50.

Remillard, Janine T., and Martha B. Bryans. "Teachers' Orientations toward Mathematics Curriculum Materials: Implications for Teacher Learning." *Journal for Research in Mathematics Education* 35 (November 2004): 352–88.

Remillard, Janine T., Beth A. Herbel-Eisenmann, and Gwendolyn M. Lloyd. *Mathematics Teachers at Work: Connecting Curriculum Materials and Classroom Instruction.* New York: Routledge, 2009.

Senk, Sharon L., and Denisse R. Thompson. *Standards-Based School Mathematics Curricula: What Are They? What Do Students Learn?* Mahwah, N.J.: Lawrence Erlbaum Associates, 2003.

Sherin, Miriam, and Corey Drake. "Curriculum Strategy Framework: Investigating Patterns in Teachers' Use of a Reform-Based Elementary Mathematics Curriculum." *Journal of Curriculum Studies* 41 (June 2009): 467–500.

Smith, Margaret Schwan, and Mary Kay Stein. "Selecting and Creating Mathematical Tasks: From Research to Practice." *Mathematics Teaching in the Middle School* 3 (February 1998): 344–50.

Tarr, James E., and Ira J. Papick. "Collaborative Efforts to Improve the Mathematical Preparation of Middle Grades Mathematics Teachers." In *The Work of Mathematics Teacher Educators: Exchanging Ideas for Effective Practice*, AMTE Monograph 1, edited by Tad Watanabe and Denisse R. Thompson, pp. 19–34. San Diego: Association of Mathematics Teacher Educators, 2004.

Van Zoest, Laura R., and Jeffrey V. Bohl. "The Role of Reform Curricular Materials in an Internship: The Case of Alice and Gregory." *Journal of Mathematics Teacher Education* 5 (September 2002): 265–88.

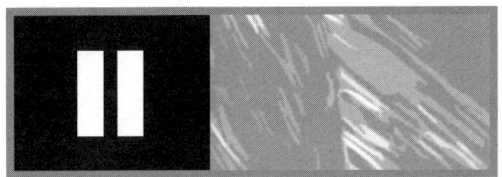

Principles for School Mathematics

CHAPTER 2

The Equity Principle: Developing Preservice Secondary School Mathematics Teachers' Conceptions of Equity—an Application of Standards-Based Curriculum Materials

Vanessa R. Pitts Bannister
Gina J. Mariano

> Excellence in mathematics education requires equity—high expectations and strong support for all students.... Expectations must be raised—mathematics can and must be learned by *all* students (NCTM 2000, pp. 12–13).

THE EQUITY Principle (NCTM 2000) encourages teachers to think about mathematics instruction in new ways. In particular, NCTM's vision of equity challenges teachers to raise expectations for all students' mathematical learning and to offer instruction that responds to students' prior knowledge and academic strengths. As mathematics teacher educators, we face difficult questions about how to prepare teachers not only to understand but also to carry out the concepts behind the Equity Principle. For preservice teachers (PTs) to enact this vision in their future mathematics instruction, they need supported, focused experiences that allow them to identify and explore issues of equity.

In this chapter, we share examples of how we have used Standards-based curriculum materials in our teacher education courses to challenge PTs' conceptions of equity. Activity 2.1 has PTs examine units from Standards-based mathematics curriculum materials and answer questions that promote the concept of high expectations. In activity 2.2, PTs develop their understandings of differentiation as it relates to the Equity Principle and identify strategies to address diverse students' needs by using mathematics curriculum materials.

The work described in this chapter was supported in part by the National Science Foundation (grant no. 0536678). Any opinions, findings, conclusions, or recommendations expressed herein are those of the authors and do not necessarily reflect the views of the National Science Foundation.

Using Standards-Based Curriculum Materials to Extend Preservice Teachers' Conceptions of Equity

In our teacher education courses, we have designed activities surrounding using Standards-based curriculum materials to help PTs develop and explore their understandings of equity. In particular, we mean the activities to supply a context and experience for PTs to consider the pedagogical implications of the Equity Principle. Some of the activities include the following:

- Supporting PTs as they evaluate how a particular curriculum unit exhibits high expectations

- Sorting tasks and problems from curriculum materials according to their levels of cognitive demand (e.g., Arbaugh and Brown 2004)

- Identifying ways to differentiate within a particular curricular activity for diverse learners, including students with special needs

- Making focused observations of the materials and social resources available to support using Standards-based curriculum materials in a variety of schools (e.g., urban and rural schools, and high- and low-achieving schools serving low-income and minority students)

We next elaborate on two of these activities.

Identifying High Expectations in a Curricular Activity

Teachers' expectations have significant impact on learning and teaching practices (Knapp 1995; Oakes 1985). These expectations derive from knowledge and beliefs about who their students are and what they can do (Solomon, Battistich, and Hom 1996). The Standards thus emphasize high expectations for all students and suggest that "high expectations can be achieved in part with instructional programs that are interesting for students and help them see the importance and utility of continued mathematical study for their own futures" (NCTM 2000, p. 13). Teachers aiming for this goal may benefit from structured opportunities to identify and evaluate ways to exhibit facets of high expectations.

Using activity 2.1, we had PTs evaluate lessons and units of the Connected Mathematics Project (CMP; Lappan et al. 2005). Before doing activity 2.1, PTs read "High Expectations: A 'How' of Achieving Equitable Mathematics Classrooms" (Jamar and Pitts 2005). This article offers manifestations of high expectations as demonstrated by Mr. Lee, an urban middle school mathematics teacher. Mr. Lee's examples portray the following aspects of high expectations:

- Use students' prior knowledge as building blocks to new knowledge, which let students know that they already have the foundation needed to learn.

- Expect students to be active participants in their own learning, which lets students know that they are responsible for their own learning.
- Give students opportunities to *understand* concepts before learning rules, which lets students know that they *can* understand the content and that it is understandable.

As PTs used ideas from Jamar and Pitts (2005) and other pertinent readings to examine the notion of high expectations in Standards-based curriculum materials, they created a range of conceptions of equity and diversity. Whereas some PTs showed narrow conceptions of equity, others seemed to develop a deeper understanding of high expectations that the curriculum materials exhibited.

Consider, for example, PTs' evaluations of the investigation "Writing Expressions for Surface Area" from CMP's "Say It with Symbols" (Lappan et al. 2005) unit. This investigation focuses on developing middle grades students' understandings of an important measurement concept (CCSSI 2010; NCTM 2000).

> I did not see many, if any at all, examples of the lesson addressing the issues of equity/diversity within the classroom.

> This unit did not seem to address directly many issues of equity and/or diversity within the classroom. The teacher's manual did suggest in the margin that linguistically diverse students try representing ideas as "rebuses" to help them obtain a better grasp on certain concepts. For example, given the word "commutative," a student could use the rebus $(2 + 3 = 3 + 2)$. Although this could help linguistically diverse students understand the idea better, one generally doesn't teach about the commutative property *without* giving examples anyway, so this seemed like a very staged attempt at aiding teachers in addressing diversity issues.

We considered the preceding responses *narrow* because they lack interpretations of teaching and learning. More specifically, the responses did not clearly show how the PTs situated teaching and learning with respect to equity and diversity issues or how the PTs' perceptions of the investigation might have influenced how they will structure learning opportunities for *all* students. We also considered the following responses narrow because they highlight a correspondence between addressing diversity issues and incorporating ethnic or foreign names:

> Problems incorporate diverse names, such as Kwang-Hee, which helps minority students identify with the questions.

> There are many foreign-sounding names such as Kiran and Masako. I assume the point of this is to include students who may feel like they are the only nonwhite person in the class.

Although some PTs initially showed narrow views of equity, we found that others seemed to understand and appreciate concepts such as *high expectations* and *strong support,* on which the Equity Principle centers. The following excerpts show how the PTs identified examples of high expectations. One PT gave the following response about the "Writing Expressions for Surface Area" investigation:

> It acknowledges the existence and importance of prior knowledge, as well as the importance and power of mathematical intuition. Students are learning this mathematics in the context of their lives and experiences. Algebraic concepts and some symbolic reasoning have been touched upon starting in sixth grade; so, when students get to this lesson, they already have some idea as to how symbolic representations operate. . . . It allows for considerable flexibility in how students approach the problems of finding surface areas of a stack of staggered rods. Students are free to use tables, pictures, or the manipulatives (Cuisenaire rods) to help them develop algebraic expressions modeling the situation.

The PT identified ways that the investigation highlights aspects of high expectations and presented findings parallel to those of Jamar and Pitts (2005): value students' prior knowledge, varied approaches, and constructions of knowledge. Another PT also identified prior knowledge as an equity issue and offered the following recommendation:

> As far as issues of equity and diversity, it may be worthy to note that this lesson sometimes requires students to retrieve prior knowledge that they were "supposed" to learn in a previous class or year of school. However, as a teacher, it is not always safe to assume that your students know something. If you do, you may lose some students who will be confused throughout the lesson, because they may have been sick last year when that lesson was taught, or maybe they just didn't understand it the first time, and need a different approach. It is important to make sure that all students are on the "the same page" and refresh them or remind them of a certainty that is needed to complete the lesson.

As they identified facets of high expectations, many PTs began to use the manifestation of high expectations as a criterion for evaluating Standards-based curriculum materials. One PT wrote the following:

> This lesson is definitely one that I would consider using in my classroom. Looking back, surface area for me was, "Here are the formulas for surface area of different surfaces. Here's a worksheet with different surfaces and their dimensions. Find the surface area of each of them." For the most part, that was my mathematics education, so reading lessons like these is very intriguing to me. I love how the students are expected to find their own equations and compare them to [those of] their peers. If the students are expected to find the equations themselves, then in the future, if they were to forget the equations, they are more likely to be able to come up with it again.

As the PTs' comments suggest, activity 2.1 offers PTs opportunities to reflect on their learning; hypothesize about the learning of students who use Standards-based materials; and, in some instances, develop appreciation for curricular activities that show high expectations for students' mathematical learning.

Understanding Differentiation in a Curricular Activity

The Standards (NCTM 2000) advocate for instructional programs that consider diverse student populations to create classrooms in which all students feel empowered to learn mathematics. Continuing our focus on the Equity Principle and Standards-based mathematics curriculum materials in our teacher education course, we used activity 2.1, part 3b, in two ways. First, it gathered information about PTs' initial conceptions of differentiation. Second, it served as an introductory activity for activity 2.2, which aims to expand PTs' conceptions in this area.

We had PTs analyze Standards-based curriculum units and propose ways in which they might alter the units to address the needs of diverse student populations. In the following excerpts, PTs expressed a broad view of equity in teaching and learning with respect to adaptation, language barriers, technology, and inclusion:

> I saw not much of a way of expanding the lesson for advanced students, or a way to condense for students that are struggling. Adaptation—spending more time on a topic before moving on.

> The unit is very wordy. This may cause some difficulty for students with language/reading deficits. [In] the teacher's manual, there is a tip for linguistically diverse classrooms, and the use of manipulatives and symbols is language-neutral, but the descriptive sections seem prohibitive to English language learners.

> There is one specific adaptation that I would like to see implemented in order for the unit to be fair for everyone, and that involves the use of calculators. While I do feel it is important to allow students the opportunity to use calculators when both the teacher and students feel it is appropriate, I can't help but wonder about those students who cannot afford to buy graphing calculators.... It is important to me that my students all have an equal opportunity to learn and grow within a mathematics context. Therefore, I feel that either the school should supply graphing calculators ... this lesson should be implemented without the use of calculators, or the teacher should use a calculator system in which examples may be presented using a calculator and screen visible to all students and with which students may take turns learning to use within the classroom. Using one of these methods would more equally include all students in the lesson/unit.

> The group-work setup would be the main place where some students may be feeling left out. However, as Boaler (2006) pointed out, teachers need to walk around and assign competence to individuals in each group who may not feel included in the lesson. With a curriculum like this, the main challenge for teachers will be to walk around and not just forget about the students. While it is the students who are doing the teaching, teachers must be readily available to assist in the learning process.

In these comments, PTs identified potential equity issues related to technology, language, and cooperative learning, as well as ways of attending to them. They also identified strategies to address the issues. As these comments suggest, using activity 2.1 can create an opportunity for teacher educators and PTs to develop and discuss possible strategies to address the needs of diverse students by using curriculum materials.

Conclusion

PTs' understandings of equity and diversity issues may affect how they use curriculum materials. Giving PTs focused work involving Standards-based curriculum materials can help them develop deeper understandings of equity and diversity issues. But, most important, the focused work can show them the many ways they can use such curriculum materials to achieve equity in mathematics classrooms (Schoenfeld 2002). Through guided interactions with Standards-based curriculum materials and consideration of how these materials relate to the Equity Principle, teacher educators better prepare future generations of mathematics teachers, who will offer fruitful mathematical opportunities to *all* students.

REFERENCES

Arbaugh, Fran, and Catherine A. Brown. "What Makes a Mathematical Task Worthwhile? Designing a Learning Tool for High School Mathematics Teachers." In *Perspectives on the Teaching of Mathematics*, 2004 Yearbook of the National Council of Teachers of Mathematics (NCTM), edited by Rheta N. Rubenstein, pp. 27–41. Reston, Va.: NCTM, 2004.

Common Core State Standards Initiative (CCSSI). *Common Core State Standards for Mathematics.* Washington, D.C.: National Governors Association Center for Best Practices and the Council of Chief State School Officers, 2010. http://www.corestandards.org.

Boaler, Jo. "How a Detracked Mathematics Approach Promoted Respect, Responsibility, and High Achievement." *Theory into Practice* 45 (Winter 2006): 40–46.

Jamar, Ido, and Vanessa R. Pitts. "High Expectations: A 'How' of Achieving Equitable Mathematics Classrooms." *Negro Educational Review* 56 (July 2005): 127–34.

Knapp, Michael S. *Teaching for Meaning in High-Poverty Classrooms.* New York: Teachers College Press, 1995.

Lappan, Glenda, James Fey, William Fitzgerald, Susan Friel, and Elizabeth Phillips. *Say It with Symbols: Making Sense of Symbols.* Englewood Cliffs, N.J.: Prentice Hall, 2005.

National Council of Teachers of Mathematics (NCTM). *Principles and Standards for School Mathematics.* Reston, Va.: NCTM, 2000.

Oakes, Jeannie. *Keeping Track: How Schools Structure Inequality.* New Haven, Conn.: Yale University Press, 1985.

Schoenfeld, Alan. "Making Mathematics Work for All Children: Issues of Standards, Testing, and Equity." *Educational Researcher* 31 (January–February 2002): 13–25.

Solomon, Daniel, Victor Battistich, and Allen Hom. "Teacher Beliefs and Practices in Schools Serving Communities that Differ in Socioeconomic Level." *Journal of Experimental Education* 64 (Summer 1996): 327–47.

Activity 2.1
Equity and Curriculum Materials: High Expectations

During this activity, you will analyze the extent to which lessons and units of a Standards-based curriculum program incorporate ideas of equity in light of the Equity Principle. After analyzing the materials, propose ways to alter the units to address the needs of diverse student populations.

1. Before analyzing curriculum materials, understanding the various facets of the Equity Principle is important. The following is a list of suggested readings:

   Boaler, Jo. "How a Detracked Mathematics Approach Promoted Respect, Responsibility, and High Achievement." *Theory into Practice* 45 (Winter 2006): 40–46.

   Hilliard, Asa III. "Do We Have the *Will* to Educate All Children?" *Educational Leadership* 45 (September 1991): 31–36.

   Jamar, Ido, and Vanessa R. Pitts. "High Expectations: A 'How' of Achieving Equitable Mathematics Classrooms." *Negro Educational Review* 56 (July 2005): 127–34.

   National Council of Teachers of Mathematics (NCTM). "The Equity Principle." In *Principles and Standards for School Mathematics*, pp. 12–14. Reston, Va.: NCTM, 2000.

   Schoenfeld, Alan. "Making Mathematics Work for All Children: Issues of Standards, Testing, and Equity." *Educational Researcher* 31 (January–February 2002): 13–25.

2. After reading the preceding articles and other readings discussing the Equity Principle and its meaning and intent, please select one unit from a Standards-based curriculum program to evaluate. Make sure to investigate all aspects of the unit, including but not limited to the following:

   - Lesson ideas and activities
   - Suggested pretests and posttests
   - Differentiated homework options
   - Notes in the teacher's guide regarding differentiation
   - Supplemental units or materials, related to differentiation, that curriculum program's publisher might contribute

3. Use the following questions to guide your analysis of the chosen curriculum unit. *Note: These questions are only a starting point. Use what you have learned from your readings on high expectations to guide your analysis.*

   a. How do you think the curriculum unit will assist you with issues of equity and diversity?

   b. What adaptations, if any, do you think are needed to address issues of equity and diversity?

Activity 2.2
Equity and Curriculum Materials: Differentiation

In this project, you will analyze the extent to which various Standards-based mathematics curriculum units offer strategies for differentiation.

1. Before analyzing curriculum materials, understanding the various facets of the term *differentiation* is important. The following is a list of suggested readings:

    a. Books or chapters on differentiation:

    - Bender, William N. "Planning for Differentiated Math Instruction." In *Differentiating Math Instruction: Strategies that Work for K–8 Classrooms*, pp. 27–46. Thousand Oaks, Calif.: Corwin Press, 2005.

    - Tomlinson, Carol Ann. *The Differentiated Classroom: Responding to the Needs of All Learners.* Upper Saddle River, N.J.: Pearson, 1999.
    Suggested readings: chapters 1, 7, and 8

    - Wormeli, Rick. *Differentiation: From Planning to Practices, Grades 6–12.* Portland, Maine: Stenhouse, 2007.
    Suggested readings: chapters 2 and 3; chapter 5, scenario 2

    b. Online resources and articles on differentiation:

    - "Differentiation Module"
    http://www.k8accesscenter.org/training_resources/differentiationmodule.asp

    - "What Is Differentiated Instruction?"
    http://www.k8accesscenter.org/training_resources/documents/Math%20Differentiation%20Brief.pdf

2. After reading several chapters or articles on differentiation, select one unit from a Standards-based curriculum program to evaluate. Make sure to investigate all aspects of the unit, including but not limited to the following:

    - Lesson ideas and activities
    - Suggested pretests and posttests
    - Differentiated homework options
    - Notes in the teacher's guide regarding differentiation
    - Supplemental units or materials, related to differentiation, that curriculum program's publisher might contribute

3. Use the following questions to guide your analysis of the chosen curriculum unit. *Note: These questions are only a starting point. Use what you have learned from your readings on differentiation to guide your analysis.*

   a. Does the unit include pretests and posttests? Does it include alternative assessment activities?

   b. Does the unit suggest alternative activities for struggling or accelerated learners?

   c. Do the materials have room for teachers to make on-the-spot decisions regarding remediation and extension activities? Explain.

   d. Do the materials suggest a variety of activities based on students' interests?

   e. Do the materials encourage students' choice among a variety of activities? If not, give examples of how you might accomplish this.

   f. Do the materials suggest a variety of activities based on students' learning profiles?

   g. Do the materials offer various product outcomes so that students may choose how to demonstrate required unit knowledge? If not, give examples of how you might accomplish this.

   h. Does the unit suggest tiered lessons or multiple entry and exit points for struggling or accelerated learners?

4. After you have answered the questions in #3, draft a letter to the curriculum committee of a local school board regarding the curriculum unit you analyzed. Your letter should do the following:

   a. Explain the extent to which the unit, as currently written, offers teachers options for differentiation. Use questions from #3 as a guide.

   b. Suggest ways that teachers might modify the unit to include multiple levels and types of differentiation for their students.

CHAPTER 3

The Curriculum Principle: Looking through Lenses—a Tripartite Examination of Textbooks

Beth A. Herbel-Eisenmann

CURRICULUM materials affect what and how teachers teach, as well as what and how students learn (Alexander and Kulikowich 1994; Begle 1979; Tobin 1987; Usiskin 1985). With the importance of curriculum materials to mathematics teaching and learning, introducing preservice teachers (PTs) to frameworks for examining curriculum materials is also important. The Curriculum Principle suggests that mathematics curriculum should be coherent, focus on important mathematics, and be well articulated across the grades (National Council of Teachers of Mathematics [NCTM] 2000). The Curriculum Principle, however, does not suggest how to examine these aspects.

In this chapter, I share three assignments that I designed and used in mathematics methods courses for PTs. Each assignment offers opportunities for PTs to analyze aspects of curriculum materials that they might not have noticed as students of mathematics, including some of the points raised in the Curriculum Principle. I also refer to aspects of the Equity Principle, because a primary goal of the third assignment is to look at curriculum materials through an equity lens. I developed the first assignment as a larger, take-home project; the second and third give PTs opportunities to revisit their understandings of the textbooks they initially reviewed during class time. In the following sections, I describe each assignment and my purposes for assigning them. I also give examples of responses I have received and share some insights about what might be valuable about the activities, particularly as they relate to aspects of the Curriculum and Equity Principles.

The Textbook Adoption Committee Project

When I taught junior high school mathematics, I was involved in two different textbook adoptions. I had no experience making such decisions, and the school districts gave no criteria or lenses for making them. I recall hearing my more experienced colleagues comment on the pictures in the book or on some of the visual representations of fractions that appeared in the book. After spending time developing the assignments I describe here, I now realize that these were fairly superficial aspects of the textbook. Textbook sales is a billion-dollar industry (Apple 1989), and adoption cycles now seem to occur less frequently because of cost-cutting measures in school districts. New and beginning teachers thus need to be better prepared to analyze and make informed decisions regarding textbook adoption. Like me earlier in my career, however, these new professionals may not have lenses for looking at textbooks. The Textbook Adoption Committee Project (activity 3.1) grew out of these experiences, observations, and concerns. This assignment's primary components are to analyze a set of curriculum materials by using a framework and curriculum standards (CCSSI 2010; NCTM 2000); to learn to make an informed decision about whether to recommend or reject a set of mathematics curriculum materials; and to write an administrator a clear, succinct, convincing letter that gives evidence for the informed decision.

After reviewing some of the frameworks for examining curriculum materials that have appeared recently in professional journals, I decided to use "Evaluating Instructional Materials" (Bernhard, Lernhardt, and Miranda-Decker 1999) because the authors wrote it for teachers, it focuses on many important aspects of teaching and learning mathematics, and it provides good questions. The framework divides the analysis into four areas: content, technology and instructional tools, assessment, and teacher support. Some specific questions in the framework focus on aspects of coherence and on the quality of mathematics being offered to students, as suggested in the Curriculum Principle. For example, some of the focus questions include the following (p. 174):

> Do the materials embed mathematics content in real-world contexts and connect mathematics to other subject areas? Are connections made among mathematics topics? . . . Are mathematical ideas clearly introduced and reinforced with examples and multiple representations, such as diagrams, graphs, and tables? . . . Do a variety of robust problem situations encourage students to explore mathematics?

The authors developed the instrument after reading a range of such tools and noticing that few had criteria for technology or assessment. The framework also includes criteria for considering issues related to equity and multicultural education. We revisit these questions later in class.

When I assign the project, I place the PTs on textbook adoption committees according to the grade-level bands they are interested in or are working in for their practicum placements. Typically, the committees comprise about three to four people, and each group receives one set of curriculum materials. The sets of materials consist of a broad range of textbooks, including some materials developed with funding from the National Science Foundation (e.g., Connected Mathematics, Mathematics in Context) and some that were not (e.g., Saxon's *Math 7/6*, Holt's *Mathematics: Exploring Your World*). Since NCTM's (2000) *Principles and Standards* suggests content and processes for the grade bands K–2, 3–5, 6–8, and 9–12, I typically assign textbooks intended for the final year in each grade band. The exception is when I select high school textbooks. I base those decisions typically on specific content such as algebra or geometry. When using integrated curriculum materials, I encourage PTs to try to see which aspects of the textbook they would consider algebraic or geometric and to select a few essential big ideas (e.g., function) to trace across the textbooks.

I have used this assignment in different courses, and I have made different modifications to make it work with different kinds of PTs. Sometimes I have asked the committees to form subcommittees, each of which works on a part of the textbook analysis. The subcommittees report back to one another, make a consensus decision, and then together write the letter to the administrator. Sometimes I request that the committees focus in-depth on one or two parts of the framework but that they comment on the others more generally. Sometimes I narrow the content down to a specific area such as "algebra" or "rational numbers." (Activity 3.1 is an example of this assignment used in a course for PTs called Algebra in K–12 Classrooms.) Each variation helps the assignment fit the course goals and offers PTs an opportunity to gain experience looking at aspects of curriculum materials.

Each committee has a week or two to analyze their assigned curriculum materials by using the framework that Bernhard, Lernhardt, and Miranda-Decker (1999) articulated. The committee members then use some class time to talk through their findings, to share examples and evidence they attended to in their analysis, and to formulate the decision and select evidence to support it. Each committee is asked to prepare a brief letter—about three single-spaced pages—to a curriculum committee or principal in which they (1) inform the committee or principal about what *should* be included in that grade band, on the basis of information from the Content and Process Standards (NCTM 2000) and the Common Core State Standards (CCSSI 2010); (2) recommend whether to adopt or reject the materials; and (3) offer a thorough, succinct, convincing evaluation of the curriculum materials. When the assignment is due, each curriculum committee briefly reports to the class about their recommendations and the evidence they drew on to develop them. One of the hardest aspects of this project for the PTs is having to give appropriate evidence for the claims they make about the curriculum materials. Some PTs have suggested that the assignment allow a longer report so that they can give more evidence. One PT suggested that the longer report be preceded by an executive summary that a committee could give to the school's principal first.

PTs have reported that, after completing the assignment, they have a much better sense of how to use specific criteria to look broadly across a set of curriculum materials. Most groups end up reporting both strengths and weaknesses of the textbooks, which can lead to a discussion about using curriculum materials as a flexible tool rather than as the only resource from which to teach. For example, one group noticed that most exercises in the textbook consisted of sets of problems that allowed students to practice skills. The group suggested that they might need additional resources, which would also add to the adoption costs, to create problem-solving experiences for students. One PT reported that this analysis helped him think about whether all aspects of the curriculum materials were strong (e.g., whether they focused primarily on symbolic manipulation or what role problem solving played in the curriculum materials) and to identify places where he thought he would need to seek out supplementary activities and resources. Another group raised concerns that their curriculum materials did not have a glossary in the back of the book. They were worried that students and parents might need that resource and suggested that they make one available. Still another group noticed that some integrated curriculum materials did not include quadratic relationships until the second course in the sequence. They suggested parents and students might need an explanation for this omission, to avoid having them unnecessarily conclude that their course was not covering what the students needed. Some groups found research about their curriculum materials and included that information in their evidence for their recommendation. Some ideas from this assignment reappear later in two other course assignments, the Types of Tasks analysis and the Model Reader analysis.

Looking Closely at Types of Tasks

Although the Textbook Adoption Committee assignment offers an opportunity to begin a fairly careful examination of textbooks, some important characteristics of textbooks could escape PTs' attention. For example, although the Textbook Adoption Framework attends to content and asks about worthwhile mathematical tasks, it does not delineate the mathematical tasks' different cognitive demands. This specific aspect of curriculum materials is important because the cognitive demand clearly influences students' opportunities to learn (Stein and Lane 1996). To attend to this aspect of mathematical tasks, PTs in my courses read about and engage with a framework about the cognitive demands of tasks (e.g., Smith and Stein 1998; Smith et al. 2004). The framework that Stein, Smith, Silver, and colleagues offer gives criteria for distinguishing among mathematical tasks of low and high cognitive demand. More specifically, it delineates between two types of low-level tasks (*memorization* and *procedures without connections*) and two types of high-level tasks (*procedures with connections* and *doing mathematics*). In particular, PTs use grade level–appropriate tasks and sort the tasks on the basis of the tasks' high- and low-level demands, as Smith et al. (2004) suggested.

After PTs seem to understand the different types of tasks, they engage in an extended, in-class activity in which they revisit the curriculum materials that they analyzed for their textbook adoption assignment. PTs receive a handout (see activity 3.2) and use the task framework to analyze the mathematical tasks in the curriculum materials more carefully. PTs have noticed that some textbooks offer primarily low-level tasks, whereas others offer primarily high-level tasks. Some PTs notice that some textbooks incorporate high-level tasks only toward the ends of the chapters or sections, and they express concern that those might go unused because of their location in the book. Other PTs raise questions about how to supplement textbooks that seem to include only one kind of task. We discuss the importance of matching learning goals with the types of tasks selected. Typically, I remind PTs that all four kinds of tasks are important and that none should be used exclusively, since all four have their place in mathematical activity. We also discuss how almost none of the PTs received tasks that would fit the *doing mathematics* category when they were students, as well as why including such tasks might be especially important. We revisit the research that showed that exposure to high-level tasks typically results in more learning gain, even in instances that reduce the task's level during implementation (Stein and Lane 1996). This set of experiences typically creates an opportunity to talk about selecting appropriate tasks for the PT's first lesson plan and about how high-level tasks tend to support the development of conceptual understanding, mathematical argumentation, and explanation in ways that lower-level tasks do not. We also read about how PTs might modify textbook problems to make the problems high level (e.g., Kabiri and Smith 2003; Kaput and Blanton 2003).

The first two examinations of textbooks that I have described focus primarily on the mathematics in which PTs are asking students to engage. Another important dimension of teaching, however, is attending to social aspects in the classroom and beyond. For example, as the Equity Principle states, "All students, regardless of their personal characteristics, backgrounds, or physical challenges, must have opportunities to study—and support to learn—mathematics" (NCTM 2000, p. 12). An important social aspect of curriculum materials for PTs to examine is whether the curriculum materials reflect their students' personal characteristics, backgrounds, physical challenges, and so on. If not, it might then be important for the teacher to modify aspects of the curriculum materials to create other opportunities for their students to see themselves reflected in the offered mathematical activities. (For an interesting example of a collective effort to modify curriculum materials to suit actual students better, see the El Barrio-Hunter College PDS Partnership Writing Collective [2009].) Because equity issues make it important that PTs consider the *specific* students they are teaching and the kinds of messages mathematics textbooks might send, we examine the curriculum materials one more time. This time, however, they focus primarily on equity and multicultural education, as they prepare a lesson for a group of children in their placement classrooms.

Looking Closely at Who the Model Reader Might Be

The third way I have PTs examine curriculum materials grows out of my research and interest in how textbooks embody beliefs and values as well as construct a "model reader" (e.g., Love and Pimm 1996). An obstacle related to writing curriculum materials is that the authors can never know who their *actual* readers are. Instead, they can only imagine who their *model* reader might be. Yet, for actual readers to see themselves as doers of mathematics, they must see themselves reflected in the textbook's model reader.

A thread through my methods courses focuses on equity, multicultural education, and social justice. For example, PTs read and discuss information about students' "funds of knowledge" (Rosebery, McIntyre, and Gonzalez 2001), democracy and equity in teaching mathematics (Ball, Goffney, and Bass 2005), unintentional bias in teaching mathematics (Wickett 1997), and teaching for social justice (Gutstein and Peterson 2005). They also read articles about different ways to support students who might struggle in mathematics because of learning difficulties. Most discussions before this in-class activity focus more on mathematics teaching and learning than on written curriculum materials. When we turn these lenses on textbooks, I have found it useful to have students think about who the textbook's model reader might be.

When authors write curriculum materials, they must have a model reader in mind and make assumptions about who the model reader is and what he or she is interested in, looks like, and has access to. By looking carefully at images, contexts used in problems, and learning activities, PTs can begin to think about who that textbook's model reader might be. We first look generally at a chapter or two in the book with the following questions in mind: Who is this textbook's model reader? What kinds of things does this model reader like to do? What kinds of resources might this model reader have? What kinds of occupations might he or she have access to? What kinds of resources might the model reader have available? PTs usually notice that textbooks include a fairly wide range of people, among them males, females, and people of various races and ethnicities. One group found two children who were in wheelchairs in one textbook, but others did not. No one found a child wearing a scarf, but some PTs found children wearing a Jewish kippa. Many were struck by how much the materials assumed that children liked candy and were interested in spending money. Activities that they noticed in the textbooks included riding bikes, playing instruments, reading books, and playing on a computer. Some reported surprise that textbook authors assumed everyone had a computer, but no one commented that some children might not own a bicycle. Some noticed a lot of rural references to farms and ranches; others noticed more general contexts.

We also read about "hidden messages" (Maxwell 1988), ways in which people's beliefs and values might be embedded in how mathematics problems are framed. PTs are often surprised by what they consider to be overtly value-laden problems in Maxwell's article. Some actors in the problems in Maxwell's article include peasants working in a field and guerrilla fighters in a problem from Mozambique; Christians and Turks in a problem from the United States; freedom fighters, soldiers, and civilians in a problem from Tanzania; and starving families and landlords in a problem from China. The PTs then return to the textbooks to see whether they notice anything different in the contexts from what they might have before when describing the model reader. PTs often give few examples at this point, although some may return to the references to spending money. I typically share the following two problems, taken from Gutstein and Peterson (2005, p. 6):

1. A group of youth aged 14, 15, and 16 go to the store. Candy bars are on sale for 43 cents each. They buy a total of 12 candy bars. How much do they spend, not including tax?

2. Factory workers aged 14, 15, and 16 in Honduras make McKids children's clothing for Wal-Mart. Each worker earns 43 cents an hour and works a 14-hour shift each day. How much does each worker make in one day, excluding any fees deducted by employers?

Although most PTs contend that the latter problem is politically charged, we spend time talking about what values are embedded in each problem. I point out that just because the first feels normal or typical does not mean that it does not have its own embedded values. As Gutstein and Peterson point out, the first problem hides a subtext of consumerism and unhealthy eating habits.

Many PTs report that this activity exposes that mathematics, and mathematics teaching and learning, are not value free. Some have also stated that this discussion's primary points connect with similar ideas that they consider in their literacy, science, and social studies methods courses. Recently, one student asked me why we had not been solving problems like the McKids problem all semester. She believed that this discussion helped her understand better how our program's mission statement, which emphasizes social justice, connected to mathematics teaching and learning.

Conclusion

Here I shared three assignments that I developed to engage PTs in analyzing curriculum materials. The three assignments help PTs grapple with issues raised in the Curriculum Principle and, to a lesser extent, the Equity Principle. Although the Textbook Adoption Committee assignment is the most comprehensive and includes aspects that attend to coherence and important mathematics, subtle features—important for PTs to attend to but that they might miss in the assignment—exist in curriculum materials. Two important features are ones that can help them focus on the tasks' cognitive demand and on who the model reader of the textbook might be. The two in-class activities that I developed allow PTs to revisit the textbooks they reviewed for the Textbook Adoption Committee assignment. Through these activities, PTs get a more nuanced understanding of tasks and the model reader, which allows me to work toward helping PTs engage all students in authentic, quality mathematics.

References

Alexander, Patricia A., and Jonna M. Kulikowich. "Learning from Physics Text: A Synthesis of Recent Research." *Journal of Research in Science Teaching* 31 (November 1994): 895–911.

Apple, Michael. *Teachers and Texts: A Political Economy of Class and Gender Relations in Education.* New York: Routledge, 1989.

Ball, Deborah L., Imani M. Goffney, and Hyman Bass. "Guest Editorial: The Role of Mathematics Instruction in Building a Socially Just and Diverse Democracy." *Mathematics Educator* 15, no. 1 (2005): 2–6.

Begle, Edward G. *Critical Variables in Mathematics Education: Findings from a Survey of the Empirical Literature.* Washington, D.C.: Mathematical Association of America, 1979.

Bernhard, Jamal, Melissa M. Lernhardt, and Rose Miranda-Decker. "Evaluating Instructional Materials." *Mathematics Teaching in the Middle School* 5 (November 1999): 174–78.

Common Core State Standards Initiative (CCSSI). *Common Core State Standards for Mathematics.* Washington, D.C.: National Governors Association Center for Best Practices and the Council of Chief State School Officers, 2010. http://www.corestandards.org.

El Barrio-Hunter College PDS Partnership Writing Collective. "On the Unique Relationship between Teacher Research and Commercial Mathematics Curriculum Development." In *Mathematics Teachers at Work: Connecting Curriculum Materials and Classroom Instruction,* edited by Janine T. Remillard, Beth A. Herbel-Eisenmann, and Gwendolyn M. Lloyd, pp. 118–33. New York: Routledge, 2009.

Gutstein, Eric, and Bob Peterson. *Rethinking Mathematics: Teaching Social Justice by the Numbers.* Milwaukee, Wis.: Rethinking Schools, 2005.

Kabiri, Mary S. and Nancy L. Smith. "Turning Traditional Textbook Problems into Open-Ended Problems." *Teaching Children Mathematics* 9 (November 2003): 186–92.

Kaput, James J., and Maria L. Blanton. "Developing Elementary Teachers' 'Algebra Eyes and Ears.'" *Teaching Children Mathematics* 10 (October 2003): 70–77.

Love, Eric, and David Pimm. "'This Is So': A Text on Texts." In *International Handbook of Mathematics,* part 1, edited by Alan J. Bishop, Kenneth Clements, Christine Keitel, Jeremy Kilpatrick, and Collette Laborde, pp. 371–409. Boston: Kluwer Academic Publishers, 1996.

Maxwell, Jenny. "Hidden Messages." In *Mathematics, Teachers, and Children,* edited by David Pimm, pp. 118–21. London: Hodder and Stoughton, 1988.

National Council of Teachers of Mathematics (NCTM). *Principles and Standards for School Mathematics.* Reston, Va.: NCTM, 2000.

Rosebery, Anne, Ellen McIntyre, and Norma Gonzalez. "Connecting Students' Cultures to Instruction." In *Classroom Diversity: Connecting Curriculum to Students' Lives,* edited by Ellen McIntyre, Anne Rosebery, and Norma Gonzalez, pp. 1–13. Portsmouth, N.H.: Heinemann, 2001.

Smith, Margaret S., and Mary K. Stein. "Selecting and Creating Mathematical Tasks: From Research to Practice." *Mathematics Teaching in the Middle School* 3 (February 1998): 344–50.

Smith, Margaret S., Mary K. Stein, Fran Arbaugh, Catherine Brown, and Jennifer Mossgrove. "Characterizing the Cognitive Demands of Mathematical Tasks: A Task-Sorting Activity." In *Professional Development Guidebook for Perspectives on the Teaching of Mathematics: Companion to the Sixty-sixth Yearbook*, edited by Rheta N. Rubenstein, pp. 45–72. Reston, Va.: National Council of Teachers of Mathematics, 2004.

Stein, Mary Kay, and Suzanne Lane. "Instructional Tasks and the Development of Student Capacity to Think and Reason: An Analysis of the Relationship between Teaching and Learning in a Reform Mathematics Project." *Educational Research and Evaluation* 2, no. 1 (1996): 50–80. doi:10.1080/1380361960020103.

Tobin, Kenneth. "Forces Which Shape the Implemented Curriculum in High School Science and Mathematics." *Teaching and Teacher Education* 3, no. 4 (1987): 287–98.

Usiskin, Zalman. "We Need Another Revolution in Secondary School Mathematics." In *The Secondary School Mathematics Curriculum*, 1985 Yearbook of the National Council of Teachers of Mathematics (NCTM), edited by Christian R. Hirsch, pp. 1–21. Reston, Va.: NCTM, 1985.

Wickett, Maryann S. "Uncovering Bias in the Classroom: A Personal Journey." In *Multicultural and Gender Equity in the Mathematics Classroom: The Gift of Diversity*, 1997 Yearbook of the National Council of Teachers of Mathematics (NCTM), edited by Janet Trentacosta, pp. 102–6. Reston, Va.: NCTM, 1997.

Activity 3.1
Example Project from a Methods Course Focused on Algebra

Textbook Adoption Committee Project

Each small group has been assigned a set of mathematics curriculum materials. If you were serving on a curriculum committee in your district, it would be your charge to decide whether the district should adopt these curriculum materials for you to use over the next five years. Do the following to help you make a decision:

1. Reread the algebra strand from the *Principles and Standards* (NCTM 2000) and the *Common Core State Standards* (CCSSI 2010) for the grade band that applies to your curriculum materials.

2. Read and consider the "Criteria for Evaluating Content" and "Criteria for Evaluating Assessment" by Bernhard, Lernhardt, and Miranda-Decker (1999).

3. Look through the curriculum materials to determine whether they include NCTM's and CCSSI's recommended algebraic ideas. Use the criteria in the Bernhard, Lernhardt, and Miranda-Decker (1999) article to examine *the algebraic ideas, the ideas' organization and structure, what students are asked to do,* and *how students are being assessed on the algebraic ideas.*

4. Write a three-page, single-spaced letter to your school principal that

    a. informs the principal about what the NCTM Standards and the *Common Core State Standards* recommend for algebra content for this grade band;

    b. states that you either recommend or reject the curriculum materials for adoption at your grade level; and

    c. gives **evidence** and **examples** from the curriculum materials to support your recommendation or rejection, on the basis of the *content* and *processes* you identified in your analysis.

Activity 3.2
Analyzing Tasks' Cognitive Demand

Revisit the set of curriculum materials you analyzed for the Textbook Adoption Committee project. Using the Cognitive Demand of Tasks framework we have discussed in class, investigate the following:

1. Find one of each type of task in their textbook and justify, using the definitions of each task type, that the task fits each category.
2. Describe where, when, and how one chapter offers these different kinds of tasks.
3. Determine whether the same patterns appear in other chapters.

After completing these three investigations, each group will briefly report to the class, focusing on any additional insights this new analysis offered to their evaluation of the textbook. At the end of class, each group will turn in a short reflection on any additional insights and questions they have about the curriculum materials they examined.

The Teaching Principle: Learning to Address the Teaching Principle with Standards-Based Curriculum Materials

Corey Drake
Tonia J. Land

IN THIS CHAPTER, we describe a series of activities designed to support elementary school preservice teachers (PTs) in learning to use Standards-based curriculum materials to teach mathematics according to the Teaching Principle (National Council of Teachers of Mathematics [NCTM] 2000). In recent years, we have developed these activities for our elementary mathematics methods courses to address individual elements of the Teaching Principle and facilitate discussing it more broadly. At the same time, NCTM's Principles overlap; many activities described here therefore also address aspects of each of the other five Principles. This overlap reflects not only the complexity of teachers' work but also the power of working with curriculum materials to help PTs understand and enact multiple principles. In the next section, we highlight primary elements of the Teaching Principle and then describe the developed activities.

NCTM's Teaching Principle

Principles and Standards for School Mathematics (NCTM 2000, pp. 17–19) summarizes the Teaching Principle as follows:

- Effective teaching requires knowing and understanding mathematics, students as learners, and pedagogical strategies.

  — Teachers need to understand the big ideas of mathematics and be able to represent mathematics as a coherent and connected enterprise.

  — Teachers have different styles and strategies for helping students learn particular mathematical ideas, and there is no one "right way" to teach.

Support for this work came, in part, from the National Science Foundation (grant number 0643497; Corey Drake, principal investigator). Any opinions, findings, conclusions, or recommendations expressed in this material are those of the authors and do not necessarily reflect the views of the National Science Foundation.

- Effective teaching requires a challenging and supportive classroom learning environment.
  - Worthwhile tasks alone are not sufficient for effective teaching. Teachers must also decide what aspects of a task to highlight, how to organize and orchestrate the work of the students, what questions to ask to challenge those with varied levels of expertise, and how to support students without taking over the process of thinking for them and thus eliminating the challenge.
- Effective teaching requires continually seeking improvement.
  - Using a variety of strategies, teachers should monitor students' capacity and inclination to analyze situations, frame and solve problems, and make sense of mathematical concepts and procedures.
  - Effective teaching requires continuing efforts to learn and improve. These efforts include learning about mathematics and pedagogy, benefiting from interactions with students and colleagues, and engaging in ongoing professional development and self-reflection.

Enacting teaching practices that reflect this principle can be a significant challenge for in-service teachers. It is an even greater challenge for PTs, who typically lack significant experiences teaching children and who therefore have few established teaching practices of their own on which to build or reflect. Our goal as teacher educators is to help PTs recognize and leverage curriculum materials as resources for building and enacting such teaching practices, which address important mathematics in ways that challenge and support all students.

We have organized the rest of this chapter around three types of tasks: in-class teaching, using materials to determine important mathematical and pedagogical content, and analyzing curriculum materials with a variety of criteria. In each section, we describe our goals for the tasks, how we enacted the tasks, and evidence of PTs' learning. Through and across these activities, we address all the Teaching Principle's major points, focusing particularly on understanding the big ideas of mathematics and how to highlight them, analyzing and evaluating a variety of pedagogical and curricular approaches, and considering how students might respond to instructional strategies.

In-Class Teaching Assignment

We designed the In-Class Teaching Assignment (activity 4.1) to address three primary ideas. First, we wanted PTs to practice implementing a lesson that uses the problem-solving approach called Launch–Explore–Summarize. This three-phase lesson plan model entails (1) getting students ready for solving a problem by activating prior knowledge, making sure the problem is understood, and establishing clear expectations; (2) letting students work while the teacher listens actively and gives appropriate hints and extensions; and (3) facilitating a class discussion while promoting a mathematical community, listening actively without evaluation, and summarizing main ideas (Van de Walle, Karp, and Bay-Williams 2010). As Stigler and Hiebert's (1999) research tells us, most traditional mathematics lessons are organized in the following manner: homework review, teacher's presentation of procedures, and students' individual practice of procedures. For PTs, the Launch–Explore–Summarize framework is often an unfamiliar model for mathematics teaching. It is not enough to have PTs create lesson plans structured around this framework; we believe they need to plan and implement a lesson. Second, we wanted PTs to become familiar with the features of curriculum materials that support lesson implementation and use them to enact the lesson. Third, we wanted PTs to increase their own knowledge of important mathematics content.

As listed in chapter 9 (Resources), a variety of lessons from several different curriculum programs are available. Each lesson includes teachers' and students' materials. From this collection, we chose six different lessons written for middle school teachers and students that addressed a variety of mathematical content. We used middle school materials with our elementary school PTs to engage them with mathematical content that was new and, sometimes, challenging for them as learners and as teachers. We made the materials available to groups of three or four PTs to teach to their classmates—"the students." We advised them to study the materials and know the presented problem well. This preparation included solving the problem by using a variety of methods and anticipating how others might solve it. A rubric (see activity 4.1) offered more guidance. Each lesson took approximately 20 minutes to enact. When the lesson was finished, classmates (the students) gave feedback on each element listed in the rubric. Because effective teaching practice involves reflection, we also asked PTs to reflect on the experience (see questions included with activity 4.1).

From these reflections, we learned that the In-Class Teaching Assignment is a favorite among our PTs. The following are a few quotations from our PTs reflecting on the experience:

> Facilitating the lesson in front of the class helped me to think about what it would be like in front of an actual classroom. I had to think about what kinds of problem-solving questions that students might ask me and how I would answer them—if I would give them a hint or tell them to work together, etc.

> The curriculum materials helped a lot, because it was very detailed in instructing us what to do and the steps that need to be made. I think it helped me because it demonstrated that I need to allow quite a bit of time for the students to grasp the problem and figure it out. It also helped me by reminding me that I need to know the problem really well in order to successfully answer any of the students' questions.

> Facilitating a lesson helped me to use the problem-solving approach while teaching. It prepared me to answer questions and pose questions that would help students think about the problem in numerous ways rather than from just one angle. We also discussed as a class the different strategies used to find an equation, which helped formulate ideas on how the solutions were similar and how they differed.

These excerpts and others suggest that our PTs appreciated the opportunity to teach, engage with the materials, and think critically about their lessons' facilitation. Another factor that led to the assignment's success was that the content challenged the "students." This factor allowed a more realistic simulation of a teaching experience, because the PTs were actually learning or relearning the material. One PT commented, "It was a good one to perform at a college level, because people actually had to problem-solve themselves. This wasn't something that people just knew. I liked this because it simulated a good learning environment."

A related activity that simulated using curriculum materials to teach mathematics involved having PTs imagine possible adaptations of curriculum materials and then view on video one or more classroom teachers actually enacting the materials in their classrooms. For example, PTs read a lesson from a third-grade Math Trailblazers unit (University of Illinois at Chicago 2008); listed possible adaptations of the materials; watched a classroom teacher using the materials, using video collected as part of our larger project related to teachers' use of curriculum materials; and then reflected on the adaptations and enactment decisions they noticed in the teaching episode. Similar to the in-class teaching activity described earlier (activity 4.1), the adaptation activity allowed PTs to envision themselves actually using and adapting curriculum materials in the relatively safe space of the methods classroom. We hope and conjecture that these experiences give PTs the beginning knowledge and skills needed to enact similar curriculum use practices in elementary school classrooms.

Using Materials to Determine Important Mathematical and Pedagogical Content

In this section, we describe two separate tasks that required PTs to identify important mathematical and pedagogical content through their analyses of curriculum materials. The first task is an intense analysis of the fraction lessons of two curriculum series (activity 4.2). We used the third-grade editions of Investigations (TERC 2008) and Math Trailblazers (University of Illinois at Chicago 2008). In the activity's more recent iterations, we have added a third curriculum series, Everyday Mathematics (University of Chicago School Mathematics Project 2007), but the activity's overall structure has remained the same. Our goal with this task was for PTs to compare various aspects of the fraction lessons and units in the sets of curriculum materials. We conjectured that important fraction concepts and teaching strategies would emerge from these comparisons.

We divided the class into six groups and gave each group a separate set of questions to consider. The six question sets are part of activity 4.2. After completing their analyses, each group presented their findings. Finally, the class created a Venn diagram to compare the curriculum series and serve as a culminating artifact.

This activity engaged PTs as they worked in their small groups, and it prompted substantial discussion. For instance, one group discussed not liking that the Investigations series specified brownie manipulatives for students to use when solving the following problem:

> Imagine that there are seven brownies to share equally among four people. About how many brownies do you think each person will get?

They believed the manipulatives directed students toward a certain solution strategy. However, once they saw the different ways students might solve the problem presented in the curriculum guide ($1 + \frac{1}{2} + \frac{1}{4}$, $\frac{7}{4}$, and $1\frac{3}{4}$), they changed their minds.

Figure 4.1 offers an example of a Venn diagram that this analysis process can help create. PTs often suggest that the ideas that overlap at the two circles' intersection are possibly the most important fraction concepts and most effective representations and contexts for teaching fractions. This suggestion allows us, as instructors, to discuss how teachers might identify the primary mathematical ideas and representations in a lesson, set of lessons, or unit. Identifying primary ideas can then, in turn, guide teachers as they make decisions about how and when to enact and adapt particular elements of the curriculum materials.

The second task in this section is a final exam item asking PTs to analyze two consecutive lessons from the Everyday Mathematics program (University of Chicago School Mathematics Project 2007):

> This exam has, as attachments, two consecutive lessons from the first-grade Everyday Mathematics.
>
> 1. List at least three primary understandings about measurement that students might develop through these activities. For each understanding, explain which specific parts of the lesson or lessons would help students develop that understanding.
>
> 2. What activity might you do next? Why? Use 2–3 sentences to describe the activity and explain its rationale.

Our goal with these questions was to assess whether PTs could identify the big ideas in the curriculum materials. We used this task to help us understand whether—after a semester of examining curriculum materials in a variety of ways—PTs could distinguish between the *big conceptual ideas* of a lesson and *the procedural steps* of a lesson. In other work (e.g., Sherin and Drake [2009]), we have found that most materials designers do not make the big ideas explicit and instead rely on teachers to identify the important mathematics through their reading of step-by-step procedures. In analyzing responses to this exam question, we found that, similarly, few PTs could identify primary mathematical ideas and the elements of the curriculum materials associated with those ideas. However, we believe that most PTs did begin to appreciate that curriculum materials include big mathematical ideas as well as step-by-step procedures.

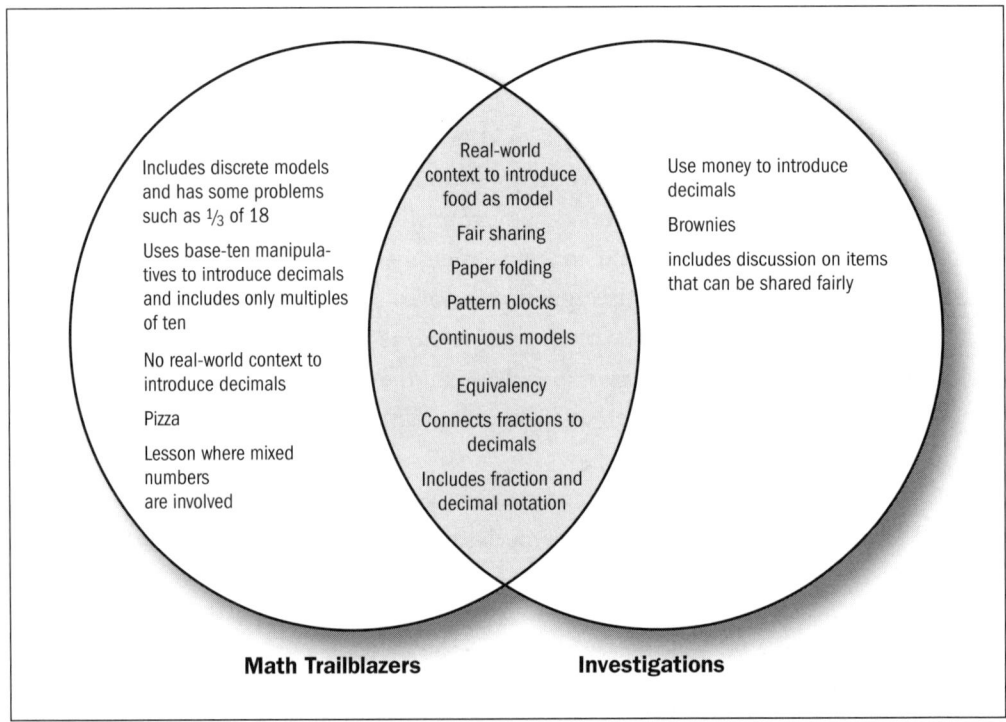

Fig. 4.1. Example of Venn diagram resulting from activity 4.2

Analyzing Curriculum Materials According to Criteria

The third type of task related to using curriculum materials has PTs analyze curriculum materials against a set of criteria (activity 4.3). In our two years of developing, enacting, and refining activities, this task type has changed the most. Initially, we used items from *Principles and Standards for School Mathematics* (NCTM 2000) or *Curriculum Focal Points* (NCTM 2006) as criteria, although we also developed a variety of additional criteria (see the second table in activity 4.3). The *Common Core State Standards* (CCSSI 2010), which many states have adopted, could serve as alternative criteria. Our goal with this activity has always been for PTs to develop their own evaluations of the materials and, perhaps more important, to understand that the process of using curriculum materials is a negotiation. In other words, when teaching, teachers are typically responsible for addressing one or more sets of standards (district benchmarks, state standards, and national standards) by using a prescribed set of curriculum materials. However, the curriculum materials often are not closely aligned with the standards and may even contradict them. Identifying and understanding these alignments and misalignments is the first step for PTs in beginning to design instruction that uses curriculum materials as a resource for addressing standards.

In table 4.1, we re-create one PT's analysis of two curriculum series that used the Focal Points related to measurement as criteria. From this activity, it seems that the PT not only gained a deeper understanding of the Focal Points as she compared them to the activities and lessons in the curriculum series, but she also developed an appreciation for the work of alignment that she will have to do as teacher.

Table 4.1.
One PT's analysis of curriculum series that used Curriculum Focal Points

Curriculum: Investigations Grade Level: 2	Curriculum: Everyday Math Grade Level: 4
Primary Activity: Comparing 2 lengths Identifying equal lengths Focal Point: Linear measurement (second grade)	Primary Activity: Identify and draw line segments, rays, and so on Focal Point: Understand measurement (second grade)
Primary Activity: Mirror symmetry Focal Point: Decomposing and composing geometric shapes (first grade) Primary Activity: Measuring using body parts versus standard units Focal Point: Ordering objects by measurable attributes (kindergarten or first grade)	Primary Activity: Measure line segments to nearest centimeter Focal Point: Understand standard units of measurements (second grade) Primary Activity: Estimate weight with and without tools Focal Point: Appropriate units/tools for estimation (fifth grade)

Table 4.1.—*Continued*

Curriculum: Investigations Grade Level: 2	Curriculum: Everyday Math Grade Level: 4
Primary Activity: Combining shapes 2-D/3-D Focal Point: Geometry—Describing shapes and space (kindergarten)	Primary Activity: Use multiplication to solve volume problems Focal Point: Volume (fifth grade)
Primary Activity: Sorting shapes by number of sides Focal Point: Geometry—Describing shapes and space (kindergarten)	Primary Activity: Area of base and surface area of rectangular prism Focal Point: Analyzing properties of 3-D shapes (fifth grade)
Summary: Most are *below* grade level.	Summary: Most seem to be *above* grade level?

In more recent iterations of this activity, we have made two important changes. First, in addition to using state standards, *Principles and Standards for School Mathematics* (NCTM 2000), and *Curriculum Focal Points* (NCTM 2006) as criteria for analyzing curriculum materials, we also ask PTs to analyze materials with respect to important ideas in learning and teaching mathematics, including the Cognitively Guided Instruction Problem Type framework (Carpenter et al. 1999) and tasks' cognitive demand (Smith and Stein 1998). This change acknowledges that teachers negotiate and adapt curriculum materials to align with particular approaches to teaching better as well as to address specific state or national standards (Drake 2010).

Second, although we continue to emphasize the curriculum unit as the focus of analysis, we now ask PTs to attend to how activities develop across a unit. For example, we have PTs examine each student page from Grade 3, Unit 1 of Investigations (TERC 2008) and identify the problem types, number choices, problem context, and cognitive demand. On the basis of this information, PTs then answer a series of questions (see the end of activity 4.3) that help summarize the unit's overall goals and how they expect children's mathematical thinking to develop over the course of the unit. Our most recent experience with this activity suggests that it not only solidifies PTs' understandings of the problem type framework and cognitive demand but also contributes a productive context for discussing concepts such as curriculum vision (Drake and Sherin 2009). The activity equips PTs with a tool (i.e., the series of questions) for examining *any* new curriculum unit, understanding children's potential trajectories through the unit, and using this information to guide daily planning and interactions with children.

Using Curriculum Materials to Teach According to the Teaching Principle

In summary, we have designed a variety of activities for methods classes to help PTs develop the skills and knowledge necessary to use curriculum materials productively. Across the semester, we want PTs to begin to understand three themes related to the Teaching Principle through engaging with these activities:

1. Curriculum materials can help teachers identify the big ideas related to particular mathematical content.

2. Curriculum materials can help teachers establish learning environments that challenge and support all students.

3. Curriculum materials can be valuable resources for teachers, but teachers must often examine materials carefully and adapt them to meet students', teachers', and the curriculum's needs (i.e., its goals and objectives).

For us, this last point deserves special emphasis. We want PTs to understand that using curriculum materials productively will often involve adapting and perhaps even supplementing the materials to support students' achievement of standards. At the same time, recognizing that certain adaptations can enhance the materials' usefulness, whereas others detract from materials' value, is important. PTs must learn to differentiate between the two types.

We believe that, through the activities described in this chapter, PTs begin to understand the importance of designing and enacting productive adaptations. For example, during the In-Class Teaching Assignment and in the subsequent reflection piece, we saw evidence of PTs' adapting the curriculum materials. One group of PTs enacting the tiling pool problem (see chapter 9, Resources) made considerable adjustments to the task. The teachers' pages directed them to present a model of a square pool with sides of length five units. Instead, they presented the task within a context (Johnny has a job tiling pools for the summer) and asked their classmates to find the number of tiles for a square pool with length 1, 2, and 3 units, and finally to make a generalization. After their classmates found the number of tiles for each length, the PTs facilitated a discussion in which classmates shared their strategies. The discussion proved useful because many were having problems understanding why the number of tiles needed for the first pool was eight instead of nine.

Participation in the methods class activities is only one aspect of the PTs' experiences in teacher education. Along with the methods course, our PTs participate in a practicum experience working with cooperating teachers in local elementary schools. One cohort of PTs completed this practicum experience in a school that had recently adopted Everyday Mathematics (University of Chicago School Mathematics Project 2007). Through this experience, the PTs acquired ideas about curriculum use that often reinforced the messages related to curriculum adaptation, as represented in the following PTs' reflections:

> From my first cooperating teacher, I saw that she had a very easy time with adjusting to the curriculum and used all aspects of it to her advantage.... When I teach mathematics, I think I will take a lot from my first placement teacher. I want math to be relaxed and flexible, and [I want] to utilize all tools given to me instead of simply drilling or doing workbooks.

> [The teacher] said that Everyday Math included many games that the children really like. My teacher pretty much taught straight out of the book and did a lot of worksheets from Everyday Math. I taught a lesson from Everyday Math, and I found out it's really important to change a lot of things to fit the kids. The lesson was pretty fun and the kids learned a lot.

> I ended up creating my own lesson when I taught. When using my own math curriculum materials, I have learned and will practice that you don't have to stick to the materials all the time and you can spice it up a bit.

Through these reflections, PTs communicate their interest in using curriculum materials as tools or resources and indicate that they expect to be able to use these tools both flexibly and in ways that make mathematics fun for children. To follow this interest, teacher educators need to consider how they can help PTs adapt and supplement curriculum materials to design lessons that are fun and that capitalize on the curriculum materials' power and potential as mathematical resources.

To accomplish this goal, we have developed and implemented the activities described in this chapter. We will continue to refine these activities and develop new ones. We believe it important to view these activities not only as stand-alone tasks for use at any point in a methods course but also as part of a trajectory of PTs' and teachers' learning about curriculum materials. This trajectory begins with PTs as mathematical learners, with their early experiences with and conceptions of curriculum material use. The trajectory continues through their teaching careers, as they experience new materials and different ways to use them. As elementary mathematics methods instructors, we are most interested in carefully designing the portion of the trajectory that takes place during the methods course. Which activities should come first for PTs, and which activities will make most sense later? Do PTs need to use materials before they can critique them? What kinds of activities, and in what order, will most successfully launch PTs into being able to use curriculum materials productively as beginning teachers? Given the considerable investment of schools and districts in curriculum materials as resources for teachers, we must prepare PTs to use these materials productively so that they can support all students' learning.

References

Carpenter, Thomas P., Elizabeth Fennema, Megan Loef Franke, Linda Levi, and Susan B. Empson. *Children's Mathematics: Cognitively Guided Instruction.* Portsmouth, N.H.: Heinemann, 1999.

Common Core State Standards Initiative (CCSSI). *Common Core State Standards for Mathematics.* Washington, D.C.: National Governors Association Center for Best Practices and the Council of Chief State School Officers, 2010. http://www.corestandards.org.

Drake, Corey. "Understanding Teachers' Strategies for Supplementing Curriculum Materials." In *Mathematics Curriculum: Issues, Trends, and Future Directions,* 2010 Yearbook of the National Council of Teachers of Mathematics (NCTM), edited by Barbara J. Reys and Robert E. Reys, pp. 277–88. Reston, Va.: NCTM, 2010.

Drake, Corey, and Miriam Sherin. "Developing Curriculum Vision and Trust: Changes in Teachers' Curriculum Strategies." In *Mathematics Teachers at Work: Connecting Curriculum Materials and Classroom Instruction,* edited by Janine T. Remillard, Beth A. Herbel-Eisenmann, and Gwendolyn M. Lloyd, pp. 321–37. New York: Routledge, 2009.

National Council of Teachers of Mathematics (NCTM). *Principles and Standards for School Mathematics.* Reston, Va.: NCTM, 2000.

———. *Curriculum Focal Points for Prekindergarten through Grade 8 Mathematics: A Quest for Coherence.* Reston, Va.: NCTM, 2006.

Sherin, Miriam, and Corey Drake. "Curriculum Strategy Framework: Investigating Patterns in Teachers' Use of a Reform-Based Elementary Mathematics Curriculum." *Journal of Curriculum Studies* 41 (June 2009): 467–500.

Smith, Margaret Schwan, and Mary Kay Stein. "Selecting and Creating Mathematical Tasks: From Research to Practice." *Mathematics Teaching in the Middle School* 3 (February 1998): 344–50.

Stigler, James W., and James Hiebert. *The Teaching Gap: Best Ideas From the World's Teachers for Improving Education in the Classroom.* New York: Free Press, 1999.

TERC. Investigations in Number, Data, and Space. Glenview, Ill.: Pearson/Scott Foresman, 2008.

University of Chicago School Mathematics Project. Everyday Mathematics. Chicago: McGraw-Hill, 2007.

University of Illinois at Chicago. Math Trailblazers. Dubuque, Ia.: Kendall-Hunt, 2008.

Van de Walle, John A., Karen S. Karp, and Jennifer M. Bay-Williams. *Elementary and Middle School Mathematics: Teaching Developmentally.* Boston: Allyn and Bacon, 2010.

Activity 4.1
In-Class Teaching

You have chosen one of six lessons to teach to your classmates, "the students." Examine the materials closely and plan a lesson accordingly. Consider how students might engage with the task along with the strategies they might use to solve. You may make any adaptations that you feel are needed. Also, have a plan in place to address challenged learners and those needing challenge. Below are the criteria on which you will be evaluated.

Category	Points out of 3	Comments
Launch—(motivating, intriguing, connects to prior knowledge, etc.)		
Explore—(students are allowed to explore mathematical concepts, teachers provide appropriate interventions when needed)		
Summarize—(discussion facilitation; several students share strategies, connections made between strategies; brings lesson to close)		
Lesson Flow—(teachers are adequately prepared, lesson is organized and flows well)		
Presentation Skills—(speaking, body language, eye contact, etc.)		

In-Class Teaching
Students' Reflection Questions

1. What do you believe went well in the lesson? What could have gone better? Explain.

2. Did the lesson plan develop strategies to solve the problem, or could students develop strategies themselves? What range of strategies did your classmates develop?

3. What was the task's level of cognitive demand (memorization, procedures without connections, procedures with connections, doing mathematics)? Did your group maintain that level during instruction? Explain.

4. Did the task have multiple entry points? Explain.

5. If you were to teach the same lesson again, what would you change? Why?

6. Did you find the curriculum materials supportive? Why or why not?

Activity 4.2
Analyzing Important Fractions Content in Two Curriculum Series

Group 1:

Look at how the two sets of third-grade materials organize fraction topics.

1. What are the similarities and differences?
2. Do they start with the same ideas about, and representations of, fractions?
3. Do they include the same ideas about, and representations of, fractions in the end?
4. What is the fraction "storyline" outlined in each set of materials? How do you know?

Group 2:

Compare the first fractions lessons in each set of curriculum materials.

1. What are the lessons' similarities?
2. What are their differences?
3. Pay particular attention to the representations, or models, of fractions in these lessons. How are they similar or different? Do you think one or the other is more accessible to third graders?
4. What are students expected to do in the next lesson in each set of curriculum materials? How do you know?

Group 3:

Compare the lessons on folding fractions.

1. What are the lessons' similarities?
2. What are their differences?
3. Pay attention to the discourse suggestions and how student-centered each lesson is intended to be. Are there any differences?
4. What do you think students will understand about fractions at the end of each lesson? What is your evidence?

Group 4:

Compare the fraction games given in each set of curriculum materials.

1. How are the games similar?

2. How are they different?

3. Do these games seem fun for third graders?

4. What do you think students will learn about fractions from playing each of these games? What is your evidence?

Group 5:

Compare the lessons introducing decimals in each set of curriculum materials.

1. How are the lessons similar?

2. How are they different?

3. What do you think students will understand about the connection between fractions and decimals at the end of each lesson? How do you know?

4. Does one lesson seem more cognitively demanding than the other? What is your evidence?

Group 6:

Compare the lessons involving fair sharing in each set of curriculum materials.

1. How are the two lessons similar?

2. How are they different?

3. Pay particular attention to the goals of each lesson. How are they similar and different?

4. What should students understand about fractions by the end of each lesson? How do you know?

Activity 4.3
Analyzing Curriculum Materials with a Variety of Criteria

Choose two different curriculum units to examine for their inclusion of NCTM's data analysis content strand. Determine whether each curriculum guide addresses all elements of the data analysis standard. In the spaces below, identify the primary activities from the curriculum guides that address each element.

Instructional programs from prekindergarten through grade 12 should enable all students to:	In grades 3–5, all students should:	Curriculum and Grade	Curriculum and Grade
Formulate questions that can be addressed with data and collect, organize, and display relevant data to answer them	Design investigations to address a question and consider how data-collection methods affect the nature of the data set		
	Collect data by using observations, surveys, and experiments		
	Represent data by using tables and graphs such as line plots, bar graphs, and line graphs		
	Recognize the differences in representing categorical and numerical data		
Select and use appropriate statistical methods to analyze data	Describe the shape and important features of a set of data and compare related data sets, with an emphasis on how the data are distributed		

Instructional programs from prekindergarten through grade 12 should enable all students to:	In grades 3–5, all students should:	Curriculum and Grade	Curriculum and Grade
Select and use appropriate statistical methods to analyze data	Use measures of center, focusing on the median, and understand what each does and does not indicate about the data set		
	Compare different representations of the same data and evaluate how well each representation shows important aspects of the data		
Develop and evaluate inferences and predictions that are based on data	Propose and justify conclusions and predictions that are based on data and design studies to investigate the conclusions or predictions further		
Understand and apply basic concepts of probability	Describe events as likely or unlikely and discuss the degree of likelihood by using such words as *certain*, *equally likely*, and *impossible*		
	Predict the probability of outcomes of simple experiments and test the predictions		
	Understand that the measure of the likelihood of an event can be represented by a number from 0 to 1		

Table adapted from Cirillo (personal communication, fall 2008).

Curriculum-Based Activities and Resources for Preservice Math Teachers

In column 2, Possible Evidence, list the kinds of things one might look for as evidence for each criterion. Use this list to guide your analyses of the materials.

Criteria	Possible evidence	Series #1	Series #2	Series #3
Alignment with Curriculum Focal Points?				
Support for students' participation in classroom discourse?				
Incorporates problem solving? When? How?				
Incorporates family, community, real-life resources?				
Support for differentiation?				
Support for assessment?				

Questions to Summarize a Unit on Curriculum Analysis

1. What do you notice about the progression of problem types?

2. What do you notice about the number choices?

3. What do you notice about the contexts of the problems and activities? Include what level of cognitive demand (memorization, procedures *without* connections, procedures *with* connections, or doing mathematics) accurately classifies most of the problems' activities and why.

4. What did you notice about which strategies students were supposed to use to solve problems? Which generated the strategies—the students or the textbook?

5. How are the curriculum designers establishing connections and coherence across the unit? List at least three different connections.

6. What would you expect to see in the next addition and subtraction unit concerning problem types and number choices? Be specific.

CHAPTER 5

The Learning Principle: Supporting the Development of Mathematical Proficiency

Fran Arbaugh

THE LEARNING PRINCIPLE states, "Students must learn mathematics with understanding, actively building new knowledge from experience and prior knowledge" (National Council of Teachers of Mathematics [NCTM] 2000, p. 20). This perspective on learning grounds itself in years of research on how students make sense of mathematics and defines knowing mathematics as something beyond memorizing facts and procedures. When teachers engage students in making sense of mathematics, the process supports a deep understanding of mathematics. Ultimately, "the kinds of experiences that teachers provide clearly play a major role in determining the extent and quality of students' learning" (NCTM 2000, p. 21).

Here I discuss the role of mathematical tasks in learning and present a research-based framework that supports teachers in examining tasks with regard to how a task supports students' learning. I then present two activities that can help preservice teachers (PTs) think about how their task choice affects students' learning, specifically as the research-based framework describes learning. These activities help mathematics teachers become more sensitive to the role of tasks in students' mathematics learning by having them consider the types of thinking that tasks require of students and what that thinking might look like in the classroom.

The Role of Mathematical Tasks in Learning

Mathematics teachers depend predominantly on their textbooks for guidance with the scope and sequencing of mathematical topics in a particular course. The teachers also look to textbooks as a source for problems with which to engage students as they learn those topics (Stein, Remillard, and Smith 2007). The types of problems that students encounter in their textbooks play a large role in what they learn mathematically as well as how they come to define what it means to learn and know mathematics (Stein and Lane 1996).

NCTM drew attention to the effect that problems have on the quality of students' learning early in the 1990s in *Professional Standards for Teaching Mathematics* (NCTM

1991). In this companion to *Curriculum and Evaluation Standards for School Mathematics* (NCTM 1989), NCTM focused the first standard on worthwhile mathematical tasks.

> **Standard 1: Worthwhile Mathematical Tasks**
> The teacher of mathematics should pose tasks that are based on—
>
> - sound and significant mathematics;
> - knowledge of students' understandings, interests, and experiences;
> - knowledge of the range of ways that diverse students learn mathematics;
>
> and that
>
> - engage students' intellect;
> - develop students' mathematical understandings and skills;
> - stimulate students to make connections and develop a coherent framework for mathematical ideas;
> - call for problem formulation, problem solving, and mathematical reasoning;
> - promote communication about mathematics;
> - represent mathematics as an ongoing human activity;
> - display sensitivity to, and draw on, students' diverse background experiences and dispositions;
> - promote the development of all students' dispositions to do mathematics. (NCTM 1991, p. 25)

NCTM continued its focus on mathematical tasks in *Principles and Standards for School Mathematics* (2000, p. 21): "When challenged with appropriately chosen tasks, students become confident in their ability to tackle difficult problems." The activities contained in this chapter focus on supporting PTs as they learn to choose and implement appropriate mathematical tasks with their future students.

Strands of Mathematical Proficiency

The authors of *Adding It Up: Helping Children Learn Mathematics* (Kilpatrick, Swafford, and Findell 2001), leaders in the field of mathematics learning, contend that learning mathematics successfully means developing *mathematical proficiency*. Having thoroughly reviewed the research literature in mathematics education and cognitive science, these authors describe mathematical proficiency as five interwoven, interdependent strands, as follows (p. 115):

- *Conceptual understanding*—comprehension of mathematical concepts, operations, and relations
- *Procedural fluency*—skill in carrying out procedures flexibly, accurately, efficiently, and appropriately

- *Strategic competence*—ability to formulate, represent, and solve mathematical problems
- *Adaptive reasoning*—capacity for logical thought, reflection, explanation, and justification
- *Productive disposition*—habitual inclination to see mathematics as sensible, useful, and worthwhile, coupled with a belief in diligence and one's own efficacy

Students' ability to understand and use mathematics requires more than memorizing facts and procedures (procedural fluency). It also requires the ability to formulate, represent, and solve problems (conceptual understanding, strategic competence, adaptive reasoning). Teachers should also support students in developing logical thought, reflection, and justification to generate the habitual inclination to see mathematics as sensible, useful, and worthwhile (productive disposition). Further, as Kilpatrick, Swafford, and Findell (2001, p. 116) state,

> These strands are not independent; they represent different aspects of a complex whole. ... The most important observation we can make here, one stressed throughout this report, is that **the five strands are interwoven and interdependent in the development of proficiency in mathematics** [emphasis in original]. Mathematical proficiency is not a one-dimensional trait, and it cannot be achieved by focusing on just one or two of the strands.

The strands of mathematical proficiency constitute a framework for what we want students to learn about mathematics: it incorporates competence, knowledge, and facility in mathematics, and it represents what teachers should strive to support in their classrooms. These strands are reflected in the Standards for Mathematical Practice of the *Common Core State Standards* (CCSSI 2010). Mathematics teacher education programs must educate their preservice and in-service mathematics teachers about the Strands of Mathematical Proficiency and how these aspects might play out in students' mathematics learning. With this goal in mind, the following sections outline two activities that I have used with PTs in my secondary methods courses.

Supporting Students' Mathematical Proficiency: The Role of Mathematical Tasks

Examining mathematical tasks through a particular lens is a powerful way for teachers to determine what students are likely to learn from using a given textbook. In chapter 3, Herbel-Eisenmann described an activity in which PTs examine the levels of cognitive demand that textbook tasks require. This chapter describes a task analysis activity in which PTs examine mathematical tasks, using the strands of mathematical proficiency as a lens. The activity's purpose is twofold: (1) to help PTs learn the five strands of mathematical proficiency and (2) to support PTs in critically examining mathematics curricular materials. As written (see activity 5.1), the activity is suitable for use with PTs. With small adaptations, it can also be powerful with in-service teachers.

For this task analysis activity, I selected and prepared a set of twelve to fifteen tasks from a variety of mathematics textbooks and curriculum materials. Teacher educators' task selection will vary according to context and course goals. For example, you may choose to consider including tasks in the set that help PTs understand a particular set of curriculum materials deeply. Many PTs have not learned mathematics from Standards-based curriculum materials and may believe that students do not learn to do mathematical procedures from these materials. You can challenge this belief by carefully choosing tasks from Standards-based curriculum materials. Another possibility is using tasks from the curriculum program that the local school district uses, so that the activity is pertinent to what the PTs see in their field classrooms.

The most important trait about the task set is that its twelve to fifteen distinct tasks all support all the strands of mathematical proficiency. Figures 5.1–3 contain a few examples of distinct tasks and the strands they support.

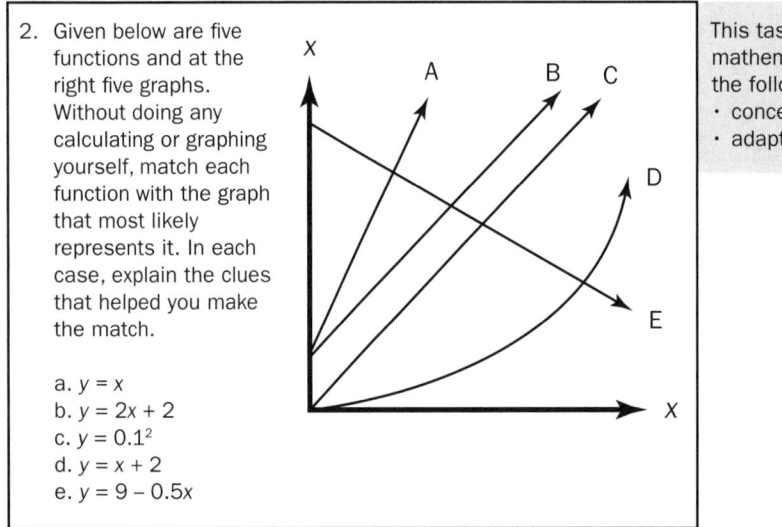

Fig. 5.1. A problem about functions (Hirsch et al. 2008, p. 168)

Before the task analysis activity, PTs should read chapter 4 from *Adding It Up: Helping Children Learn Mathematics* (Kilpatrick, Swafford, and Findell 2001). I have assigned this chapter as a course requirement, to be completed just before engaging in the task analysis in class. Using the first page of activity 5.1, PTs took notes on the chapter's salient parts, phrases, and ideas, particularly concerning each strand. PTs then came to class prepared to engage in the task analysis activity.

The Learning Principle

2. For a biology project, Nicole is invetigating how fast a particular beetle population will grow under controlled conditions. She started her experiment with 5 beetles. The next month she counted 15 beetles.

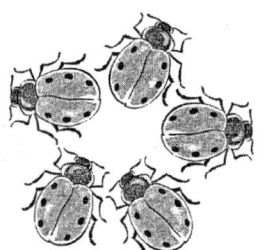

This task supports mathematical proficiency in the following strands:
- conceptual understanding
- strategic competency
- adaptive reasoning
- productive disposition

a. If the beetle population is growing linearly, how many beetles can Nicole expect to find after 2, 3, and 4 months?

b. If the beetle population is growing exponentially, how many beetles can Nicole expect to find after 2, 3, and 4 months?

c. Write an equation for the relationship between the number of beetles and the number of months if the beetle population is growing linearly.

d. Write an equation for the relationship between the number of beetles and the number of months if the beetle population is growing exponentially.

e. If the beetle population is growing linearly, how long will it take the population to reach 200?

f. If the beetle population is growing exponentially, how long will it take the population to reach 200?

Fig. 5.2. A population growth problem (Lappan et al. 2002, p. 23)

9. Find equivalent factored forms for each of these standard-form quadratic equations.

a. $x^2 + 7x + 6$
b. $x^2 + 7x + 12$
c. $x^2 + 8x + 12$
d. $x^2 + 13x + 12$
e. $x^2 + 10x + 24$
f. $x^2 + 11x + 24$
g. $x^2 + 9x + 8$
h. $x^2 + 6x$
i. $x^2 + 9x + 18$
j. $3x^2 + 18x + 24$

What general guidelines do you see for factoring expressions like these?

This task supports mathematical proficiency in the following strands:
- procedural fluency
- adaptive reasoning

Fig. 5.3. A problem about quadratic expressions (Hirsch et al. 2008, p. 338)

I organized the PTs into pairs or groups of three and then had the PTs share their notes and agree on important aspects of each mathematical proficiency strand. Using the second page of activity 5.1, the groups analyzed each task in the task set and decided which strand(s) the task supports. When each group completed analyzing the set of tasks, they compared findings with those of other groups. On tasks where the findings disagreed, we had a whole-class discussion. For example, students often disagree about whether a task supports the conceptual understanding strand. Many students look for the tasks' surface features during the analysis, such as whether a task has multiple representations (e.g., a graph and an equation). Once they see multiple representations, they "check" the conceptual understanding strand. Other students argue that *the task's requiring that students make connections among multiple representations* supports conceptual understanding. Such discussion also allows the instructor to reemphasize previous course content (e.g., readings from *Principles and Standards for School Mathematics* [NCTM 2000]) to help the PTs see threads that run through the course. This particular task analysis activity's point, however, is not to get the right answers but to begin to learn about the mathematical proficiency strands and sensitize PTs to examining tasks through those strands.

PTs generally need more than one activity built around understanding the mathematical proficiency strands to really understand and be able to use that lens as a meaningful tool to support their teaching. Several follow-up activities to this task analysis could further PTs' knowledge of, and sensitivity to, the mathematical proficiency strands:

- Analyzing a chapter of a mathematics textbook to characterize the strands of mathematical proficiency most supported by the tasks in that chapter
- Analyzing tasks used during field observations
- Searching the Internet or supplementary print materials for tasks that the PTs could use to teach specific mathematical topics and that support particular strands of mathematical proficiency

Analyzing different genres of mathematics textbooks (e.g., commercially developed and Standards-based) does not just offer PTs multiple opportunities to internalize the strands of mathematical proficiency. Such an analysis is also useful for understanding differences among different textbook series. For example, when analyzing a commercially developed algebra textbook chapter through this lens, my students determine that most of the textbook's tasks support the procedural fluency strand. They can then discuss why teachers may need to supplement the adopted textbook to support other mathematical proficiency strands and how to choose those supplements for students.

Anticipating Students' Responses to and Learning from Mathematical Tasks

Once PTs can identify the potential that tasks have for supporting students' becoming mathematically proficient, they are ready to hypothesize about students' typical and creative responses to a task, anticipate difficulties students might have with a task, and develop possible ways to address those difficulties. Anticipating students' responses is an important step in successfully orchestrating classroom discussions (Smith et al. 2009), and it is a valuable skill to enhance in PTs. In my work, PTs rarely have thought about multiple approaches to solving a problem. In fact, I have often heard comments such as "I didn't know that *my way* was different from other students'" and "I had no idea that there were so many ways to do a math problem."

Lloyd and Pitts Bannister (2010) described an activity in which they had PTs anticipate students' responses to a curricular task about surface area. In the task, from *Say It with Symbols* (Lappan et al. 2005), students must write an equation for the relationship between the number (n) of rods in a stack of staggered rods and the surface area of the stack. A stack consists of n equal-length rods that are staggered by the length of one unit rod. Figure 5.4 presents an excerpt from one PT's hypotheses about students' responses. In examining the task, the PT identified a possible misconception students might have in identifying the meaning of *variable* in the activity. The PT then described a method for redirecting such a student's attempt to find a rule without discounting that student's initial approach.

When writing an expression for the surface area of a staggered stack of n rods, students may need to use multiple variables rather than simply use n to represent the number of rods in the stack. An example of a table created by a student with this misconception might be as follows:

Number of rods	Surface area
1	$4x + 2y$
2	$6x + 6y$
3	$8x + 10y$
4	$10x + 14y$
n	???

The student is using x to represent the phrase "long sides" and y to represent the phrase "short sides" or "unit sides." The teacher could point out that they can find the *exact* surface area for one, two, three, or four rods by finding the length of the rods and working from there.

Fig. 5.4. A PT's hypotheses about students' responses to a surface area task (Lloyd and Pitts Bannister 2010, p. 332)

The example from figure 5.4 shows how PTs might use curriculum materials to hypothesize and prepare for potential learning situations. Activity 5.2 is designed as an extended project and gives PTs an opportunity to engage with the teacher's guides of Standards-based curriculum materials to extend their knowledge of the mathematical proficiency strands to hypothesize about students' responses.

Conclusions

The two activities presented in this chapter focus PTs on two very important components of planning instruction: (1) choosing tasks to support students' learning and (2) anticipating students' responses to the tasks. By engaging in activity 5.1, PTs learn to assess differences in tasks as they are written in curricular materials, focusing on how much the tasks support the five mathematical proficiency strands. Activity 5.2 then furthers the PTs' teaching knowledge by focusing them on students' anticipated responses to the tasks. Taken together, these two activities can offer PTs opportunities to learn and use valuable tools for designing and implementing the type of mathematics instruction that will help their future students learn mathematics with understanding.

References

Common Core State Standards Initiative (CCSSI). *Common Core State Standards for Mathematics.* Washington, D.C.: National Governors Association Center for Best Practices and the Council of Chief State School Officers, 2010. http://www.corestandards.org.

Hirsch, Christian R., James T. Fey, Eric W. Hart, Harold L. Schoen, Anne E. Watkins, Beth E. Ritsema, Rebecca W. Walker, Sabrina Keller, Robin Marcus, and Gail Burrill. *Contemporary Mathematics in Context, Course 1.* 2nd ed. New York: Glencoe McGraw-Hill, 2008.

Kilpatrick, Jeremy, Jane Swafford, and Bradford Findell, eds. *Adding It Up: Helping Children Learn Mathematics.* Washington, D.C.: National Academies Press, 2001.

Lappan, Glenda, James T. Fey, William M. Fitzgerald, Susan N. Friel, and Elizabeth D. Phillips. *Growing, Growing, Growing: Exponential Relationships.* Glenview, Ill.: Prentice Hall, 2002.

———. *Say It with Symbols: Making Sense of Symbols.* Glenview, Ill.: Pearson Prentice Hall, 2005.

Lloyd, Gwendolyn M. "Anticipating Student Difficulties with a Curricular Task." http://www.math.vt.edu/people/lloyd/curriculum/activities.html (accessed May 19, 2009).

Lloyd, Gwendolyn M., and Vanessa R. Pitts Bannister. "Secondary School Mathematics Curriculum Materials as Tools for Teachers' Learning." In *Mathematics Curriculum: Issues, Trends, and Future Directions,* 2010 Yearbook of the National Council of Teachers of Mathematics (NCTM), edited by Barbara J. Reys and Robert E. Reys, pp. 321–36. Reston, Va.: NCTM, 2010.

National Council of Teachers of Mathematics (NCTM). *Curriculum and Evaluation Standards for School Mathematics.* Reston, Va.: NCTM, 1989.

———. *Professional Standards for Teaching Mathematics.* Reston, Va.: NCTM, 1991.

———. *Principles and Standards for School Mathematics.* Reston, Va.: NCTM, 2000.

Smith, Margaret S., Elizabeth K. Hughes, Randi A. Engle, and Mary Kay Stein. "Orchestrating Discussions." *Mathematics Teaching in the Middle School* 14 (May 2009): 548–56.

Stein, Mary Kay, and Suzanne Lane. "Instructional Tasks and the Development of Student Capacity to Think and Reason: An Analysis of the Relationship between Teaching and Learning in a Reform Mathematics Project." *Educational Research and Evaluation* 2, no. 1 (1996): 50–80.

Stein, Mary Kay, Jeanine T. Remillard, and Margaret S. Smith. "How Curriculum Influences Student Learning." In *Second Handbook of Research on Mathematics Teaching and Learning,* edited by Frank K. Lester Jr., pp. 319–69. Charlotte, N.C.: Information Age Publishing, 2007.

Activity 5.1
Task Analysis: Supporting Mathematical Proficiency

Read chapter 4 from *Adding It Up: Helping Children Learn Mathematics* (Kilpatrick, Swafford, and Findell 2001). Using the table below, record important ideas about each strand of mathematical proficiency. Be specific enough with what you record that you do not have to go back to the chapter to understand the definition of each strand.

Strand	Important ideas
Conceptual understanding	
Procedural fluency	
Strategic competence	
Adaptive reasoning	
Productive disposition	

The Learning Principle

Analyze each task in the set to identify which mathematical proficiency strands the task supports when it is completed as it is written. Record your decisions in the table below, putting an X for each strand that the task supports. A task may support more than one strand.

	Conceptual understanding	Procedural fluency	Strategic competence	Adaptive reasoning	Productive disposition
1					
2					
3					
4					
5					
6					
7					
8					
9					
10					
11					
12					
13					
14					
15					

Activity 5.2
Anticipating Students' Responses

(Activity adapted from Lloyd [2009])

In this project, you will hypothesize about students' typical and creative responses to a task, anticipate difficulties students might have with that task, and develop possible ways to address those difficulties.

The Learning Principle stresses that "students must learn mathematics with understanding, actively building new knowledge from experience and prior knowledge" (NCTM 2000, p. 20). To attend to your future students' learning, you must develop an awareness of how mathematics understandings grow through particular experiences.

Project Guidelines

1. Review the following materials:

    a. Middle School Geometry, Grade 7: CMP's *Filling and Wrapping*
       http://matheddb.missouri.edu/showme/lesson/main.php?Project=CMP
       Download (1) teacher's pages, (2) students' pages, and (3) unit goals.

    b. Middle School Measurement, Grade 7: MathScape's *From the Group Up*
       http://matheddb.missouri.edu/showme/lesson/lesson.php?ID=MathScape74
       Download (1) teacher's pages and (2) lesson pages.

    c. High School Algebra, Grade 9: IMP's *Cookies*
       http://www.mathimp.org/curriculum/samples.html
       Download (1) classwork, (2) homework, and (3) teacher's guide.

    First, write a paragraph for each set of materials, using the five mathematical proficiency strands to characterize what students will learn through engaging with the set's tasks. Be sure to address the following questions.

    - How will students have the opportunity to enhance their mathematical proficiency across all five strands?

    - What gaps exist in these materials concerning the five mathematical proficiency strands?

    Second, as you read through these materials, pay careful attention to the parts of the teacher's guides that address students' typical responses and highlight potential areas of difficulty for students.

2. Answer the following questions about the preceding units:

   a. In your opinion, which of these units yields the most useful information about students' typical responses and potential areas of difficulty? Give specific examples, and explain why you think this information would be useful to teachers.

   b. Hypothesize which of this unit's question(s) or task(s) will be the most difficult for students. Using information in the teacher's guide as a starting point, develop possible ways that you might address students' difficulties if you were teaching this unit.

3. Choose one of the three units above. Put aside the teacher pages or teacher's guide, and focus only on the students' pages.

   a. From the students' pages, choose one task that looks interesting to you. Identify which strand(s) of mathematical proficiency the tasks supports.

   b. Hypothesize about one typical and one creative or unusual response from students to the task. In both instances, assume that students understand the task and are developing correct responses and ideas.

   c. How do your hypothesized responses in question 3b indicate that students enhanced their knowledge in one or more of the five mathematical proficiencies that you identified in question 3a? In other words, how do your responses to questions 3a and 3b correlate? Be specific.

   d. What difficulties might some students have with this particular task? Offer possible explanations for this difficulty and give sample responses to the task that might arise given this difficulty.

   e. How might you address students' difficulties? Develop at least one way to address students' misconceptions or incorrect responses.

   f. Now look at the teacher's pages or teacher's guide. What information or suggestions do they offer related to this task? Does looking at the recommendations in the teacher's guide change your ideas about questions 3a–c above? If so, how?

CHAPTER 6

The Assessment Principle: Broadening Preservice Teachers' Views of Assessment through Engagement with Curriculum Materials

Gwendolyn M. Lloyd

MOST OF THE preservice teachers (PTs) in my mathematics courses equate assessment with tests and quizzes. When they think about assessing students' understandings of mathematics, the first thing that comes to mind is a paper-and-pencil test consisting of problems similar to those that students have learned about and solved in their classwork and homework. This phenomenon is not particularly surprising; after all, PTs may not have experienced other forms of assessment in their own schooling, or they may simply not remember, or have been aware of, other strategies that their mathematics teachers used to gather information about students' understandings. Furthermore, most PTs have limited personal experience with mathematics instruction guided by students' understanding and learning.

A major challenge facing mathematics teacher educators is preparing future teachers to enact the National Council of Teachers of Mathematics (NCTM) vision of assessment (NCTM 2000, p. 22): "Assessment should support the learning of important mathematics and furnish useful information to both teachers and students." Although tests may be part of an effective program of assessment, "assessment should be more than merely a test at the end of instruction to see how students perform under certain conditions; rather, it should be an integral part of instruction that informs and guides teachers as they make instructional decisions" (p. 22). Teacher educators must identify ways to help PTs appreciate that assessment should enhance students' learning and guide teachers' instructional decisions.

This chapter shares four mathematics teacher education activities that aim to expand PTs' views of assessment and its role in teaching and learning. In the first, PTs watch a video of students working on a curricular task and develop statements, supported

---

The activities presented in this chapter were developed and implemented when the author was a faculty member at Virginia Tech. This work was supported in part by a grant from the National Science Foundation (grant number 0536678). Opinions, findings, conclusions, or recommendations herein are those of the author and do not necessarily reflect the views of the National Science Foundation.

by evidence, about students' mathematical understanding. The second requires PTs to articulate an assessment plan as part of a peer-teaching experience based on a Standards-based curriculum lesson. In the third, PTs examine the assessment resources that different curriculum programs and textbooks offer. Finally, the fourth activity engages PTs in examining samples of students' work included in the curriculum materials, as well as using and developing rubrics to assess students' understanding.

Looking for Evidence of Students' Understanding

In the *Assessment Standards*, NCTM (1995) defined assessment as "the process of gathering evidence about a student's knowledge of, ability to use, and disposition toward mathematics and of making inferences from that evidence for a variety of purposes" (p. 3). To help PTs view assessment as "gathering evidence," I developed activity 6.1. In this activity, PTs work in small groups on selected problems from a Connected Mathematics Project lesson about surface area, "Designing Packages" (Lappan et al. 1998). The PTs then watch video clips from a classroom in which students are working on and discussing the same material. As PTs watch, I encourage them to make notes about what students seem to be communicating about their understandings. I ask the PTs what the students seem to understand, and what kind of evidence they can find, by observing and listening to the students, to support their ideas about students' understanding.

PTs can find this activity challenging. It tempts them first to focus on the teacher and critique the teacher's instruction as they watch the video clips. However, as the PTs work together to develop statements about what they think the students understand about surface area, they seem to become excited about learning to observe and listen to students as an assessment technique. They had previously thought of mathematics assessment as written work and tests. I have found that PTs' familiarity with a lesson's mathematical goals and problems, developed by engaging with the lesson's problems as learners at the activity's beginning, greatly enhances their ability to attend to what the students do and say in the video clips.

For example, in one of the first problems in the "Designing Packages" lesson, students must find all the ways that they can arrange 24 cubes into a rectangular prism and then identify the arrangements with the greatest and least surface areas. This problem challenges PTs, as they work systematically to find the dimensions of the different possible prisms and then develop efficient and accurate methods for determining surface area. After solving this problem, PTs are intrigued by the variety of methods—often different from their own—that students in the video used to solve the same problem. This experience is often eye-opening as PTs begin to recognize that, by listening to students, they may not only gain information about students' thinking but also learn about connections among familiar and new strategies and approaches.

I chose the "Designing Packages" lesson because it fit with a course emphasis on surface area and volume. Video clips for many Standards-based curriculum lessons are available online, and this activity can be adapted for other lessons. Videos of sample lessons from several middle school curriculum programs are available from the Modeling Middle School Mathematics project at http://www.mmmproject.org.

Assessing Students' Understandings of Particular Mathematical Ideas

Many teacher education courses offer opportunities for PTs to develop and teach lessons to peers (i.e., other PTs). These peer-teaching experiences should include attention to assessment and its role in instruction. After all, "assessment and instruction must be integrated so that assessment becomes a routine part of the ongoing classroom activity rather than an interruption" (NCTM 2000, p. 23). When I have PTs use Standards-based curriculum materials to design instruction for their peers, I expect them to make explicit in their lesson plans how they will obtain information about their peers' knowledge and learning and how they will use that information to guide instruction. The following questions are examples of the sort I have used to guide PTs' development of their lesson plan's assessment component:

- How will you monitor changes in students' understanding during and after your lesson? What methods and tasks will you use?

- For each method or task that you include in your assessment plan, justify how the method or task will yield valuable information about students' progress.

- How do you predict that you might adjust your instructional plans, during instruction, on the basis of the information you gather about students' understanding? Give some specific examples or scenarios.

- After teaching your lesson, reflect on the assessment plan you implemented. Did it effectively monitor students' understanding during the lesson? Did it offer insights that would help you make decisions about later instruction?

Although the PTs in my classes have tended to focus their assessment programs on test and quiz items, they have also developed journal-writing prompts, self-assessment tools, and performance-based tasks.

Figure 6.1 presents two items that a group of preservice secondary school teachers used, during their peer teaching of a geometry lesson from the Core-Plus Mathematics Project (Coxford et al. 2003), to assess students' procedural and conceptual understanding of angle sums in polygons. The PTs' justification for the test item's value—how they can use the item to assess students' understanding of the lesson's main mathematical ideas—follows each problem or question.

---

1. What is the angle sum of an *n*-sided polygon? Use your answer to evaluate the sum of the angles in a triangle, a pentagon, and an octagon.

Why this is a foundational problem that assesses main points from our lesson:

Our lesson led students to discover the formula for interior angle sums of a polygon. This formula is extremely useful in geometry, and it is important that students know it and can apply it. Simply asking students to apply the formula is not necessarily useful, though, so having them demonstrate that they can apply it to various polygons ensures that they have learned the information meaningfully.

---

2. Use a visual method to show that the sum of interior angles in the polygon below is 1080 degrees. Explain the steps that you take.

Why this is a primary problem that assesses main points from our lesson:

Memorizing the formula for sums of angles may help students to solve problems quickly, but asking them to demonstrate this information in a visual manner tells us how well they actually understand why the formula makes sense.

Fig. 6.1. Two assessment items developed by a group of PTs
(Lloyd and Pitts Bannister 2010, p. 333)

The first problem in figure 6.1, which demands that students recall and apply a formula developed in the main part of the lesson, is fairly typical of the types of assessment items that PTs in my course have developed. Less common are assessment items that require students to communicate understandings verbally and visually, as in figure 6.1's second problem. In my courses, this assignment has engaged PTs in conversations about moving beyond the development of easy-to-grade problems and toward those that invite students to articulate their knowledge meaningfully and usefully. After such conversations, I have sometimes had PTs revisit their original assessment plans and make revisions to include tasks with multiple entry points and real-world contexts. I have also urged PTs to consider carefully how they might gather additional information about students' learning with methods such as observations, interviews, journals, and portfolios.

Examining Assessment Resources in Curriculum Materials

Helping PTs expand their repertoires of assessment techniques is important. As NCTM (2000) explains, "In addition to formal assessments, such as tests and quizzes, teachers should be continually gathering information about their students' progress through informal means, such as asking questions during the course of a lesson, conducting interviews with individual students, and giving writing prompts" (p. 23). With this goal in mind, I created activity 6.2, which involves examining the kinds of assessment resources that different mathematics curriculum programs offer—both Standards based and commercially developed.

When they identify the assessment resources in a teacher's guide (in question 2 of activity 6.2), most PTs in my courses have tended to pay most attention to formal assessments, such as paper-and-pencil quizzes and tests, at least initially. However, after class discussions that also identify other resources, PTs quickly realize that the assessment resources in Standards-based curriculum materials are extensive. These resources include suggested problems and writing prompts for informal assessment, questions for class discussion or individual interviews, quizzes and tests, scoring rubrics, samples of students' work, question banks, self-assessments, unit projects, and recommendations for portfolio items. As the PTs examine these resources, they can develop an appreciation for the many ways that they can monitor students' progress during instruction. In my courses for preservice elementary school teachers, the PTs become particularly excited about using journal writing and unit projects as assessment techniques in mathematics.

When PTs examine some commercially developed textbooks, which tend to offer formal assessment resources and solutions for teachers, the PTs develop an awareness of how extensive the resources included in Standards-based materials are. A follow-up activity to activity 6.2 involves having PTs develop an effective assessment plan for teaching a lesson from a commercially developed textbook. This follow-up activity can help PTs recognize when and how to adapt instructional materials. It can also reassure them that they can use effective assessment strategies even when their textbook does not contain an extensive set of resources.

Thinking about Rubrics and Samples of Students' Work

Many curriculum programs' teacher's guides give examples of students' work for various curricular tasks and assessment items, as well as rubrics to help teachers score students' work. Some programs even make students' work and rubrics available online (e.g., the Connected Mathematics Project [http://connectedmath.msu.edu/teaching/evaluate.shtml#specific]). These examples of students' work and associated scoring rubrics have tremendous potential for use by teacher educators.

In my courses for preservice elementary school teachers, I use the samples of students' work and rubrics in the *Assessment Handbook* that accompanies each grade level of the Everyday Mathematics curriculum program (UCSMP 2007). For each Everyday Mathematics unit, the *Assessment Handbook* offers a holistic scoring rubric and samples of students' work on one open-response task. For example, unit 5 of the fourth-grade curriculum focuses on multiplication algorithms and comparing large numbers—both important aspects of the number and operations strand of the elementary grades curriculum (CCSSI 2010; NCTM 2000). This task has students explain whether they can fit one million dollars in a suitcase, given the following pieces of information (UCSMP 2007, p. 178):

- You can cover a sheet of paper with six $100 bills.
- There are 500 sheets in one ream of paper.
- There are 10 reams in one carton.
- Your suitcase can hold as much as one carton of paper.

Furthermore, the million dollars is made up of $700,000 in $100 bills and the rest in $20 and $10 bills. The *Assessment Handbook* gives six samples of students' work on this task—two examples of level 4, two examples of level 3, and one example of each of levels 1 and 2—and justifications for the score applied to each sample.

After giving PTs time to complete the "Walking Away with a Million Dollars" task individually, I distribute copies of the rubric and the samples of students' work (without scores) and have the teachers work in small groups and assign a score (0–4) to each sample, as well as to their own papers. This group activity gives rise to a lively, whole-class discussion in which the PTs compare notes about their scoring of each sample of students' work. Although at first discrepancies usually will appear in how different groups score the students' work, the class typically reaches consensus after some discussion. Finally, I share with them the *Assessment Handbook*'s scoring and justifications. PTs routinely comment about how helpful it is to read the descriptions of how two very different responses from students could be scored at the same level. The scoring descriptions also help the teachers make sense of scores for students' work that seems to them to fall between levels. Because rubrics and students' sample work are provided for each Everyday Mathematics unit, this activity can be repeated many times, for different mathematical topics and grade levels.

Prior to explicit discussions and activities such as the one just described, most PTs are largely unfamiliar with rubrics or their purpose. Rubrics can help teachers not only score consistently but also recognize what mathematical understandings students are communicating in their responses to a task. Developing a rubric demands that the teacher identify the most important aspects of completing the task and clarify his or her expectations for students' responses. Because developing rubrics is difficult work, PTs may benefit from support as they attempt to develop a rubric for the first time. PTs can work together to design rubrics for assessment tasks from curriculum materials, such as "Walking Away with a Million Dollars." Once PTs develop a tentative rubric, they can attempt to use the rubrics to score a variety of students' work and, after identifying the rubrics' weaknesses, make revisions. This process will help emphasize the value of creating clear lists of indicators of understanding that can be applied to students' work. Different groups of PTs developing rubrics for the same task will proably create various rubrics, because different aspects of understanding are important to different people.

Conclusions

In this chapter, I described four curriculum-based activities that may be adapted for use in mathematics or methods courses and with preservice elementary or secondary school teachers. Common to all these activities is a goal of helping PTs view assessment as much more than easy-to-grade quizzes and tests. Standards-based curriculum programs offer tremendously useful resources to teacher educators who take seriously the need to develop PTs' views of assessment.

By observing students' work on lessons from Standards-based curriculum materials, PTs can begin to develop observational skills and recognize the importance of ongoing assessment in the instructional process. When PTs plan and teach lessons with curriculum materials, they have the opportunity to develop assessment programs that are integrated with instruction and to create tasks that allow students to communicate their knowledge meaningfully. Examining the assessment resources from different curriculum programs can allow PTs to become familiar with, and recognize the value of, the wide range of formal and informal assessment tools and techniques integral to effective mathematics instruction. And, by using and developing rubrics to score students' work on open-response tasks from Standards-based curriculum materials, PTs can come to appreciate the value of rubrics and the need for teachers to understand deeply the mathematics they are teaching and the mathematical goals of their instruction.

REFERENCES

Common Core State Standards Initiative (CCSSI). *Common Core State Standards for Mathematics.* Washington, D.C.: National Governors Association Center for Best Practices and the Council of Chief State School Officers, 2010. http://www.corestandards.org.

Coxford, Arthur F., James T. Fey, Christian R. Hirsch, Harold L. Schoen, Eric W. Hart, Brian A. Keller, and Ann E. Watkins, with Beth Ritsema and Rebecca K. Walker. *Contemporary Mathematics in Context: A Unified Approach, Courses 1–4.* Rev. ed. Columbus, Ohio: Glencoe/McGraw-Hill, 2003.

Lappan, Glenda, James Fey, William Fitzgerald, Susan Friel, and Elizabeth Phillips. *Filling and Wrapping: Three-Dimensional Measurement*. Palo Alto, Calif.: Dale Seymour Publications, 1998.

Lloyd, Gwendolyn M., and Vanessa R. Pitts Bannister. "Secondary School Mathematics Curriculum Materials as Tools for Teachers' Learning." In *Mathematics Curriculum: Issues, Trends, and Future Directions,* 2010 Yearbook of the National Council of Teachers of Mathematics (NCTM), edited by Barbara J. Reys and Robert E. Reys, pp. 321–36. Reston, Va.: NCTM, 2010.

National Council of Teachers of Mathematics (NCTM). *Assessment Standards for School Mathematics*. Reston, Va.: NCTM, 1995.

——. *Principles and Standards for School Mathematics*. Reston, Va.: NCTM, 2000.

University of Chicago School Mathematics Project (UCSMP). *Everyday Mathematics Assessment Handbook*. Chicago, Ill.: Wright Group/McGraw-Hill, 2007.

Activity 6.1
Looking for Evidence of Students' Understanding

In this activity, you will work through part of the lesson "Designing Packages," from the Connected Mathematics Project. You will then watch video clips from a classroom in which students are working on and discussing the same material. As you watch the videos, you will focus your attention on what the students seem to understand, and you will look for evidence of that understanding. In other words, you will be engaged in observing and listening to the students as a way to *assess* their understandings.

(1) Go to the Modeling Middle School Mathematics page of videos: http://mmmproject.org/video_matrix.htm and select "Designing Packages."
   Before viewing any video clips, access the teacher pages for the "Designing Packages" lesson. With a small group of your classmates, work through problems 2.1 and 2.2, and their follow-ups, and discuss the ACE questions. As you do so, try not to look at the given answers until after you discuss the questions in your group.

(2) What are the main mathematical understandings that this lesson aims to develop? If you were teaching this lesson, what "evidence of understanding" might you be looking and listening for as your students worked on and discussed these problems?

(3) Watch the video clips corresponding to this lesson. Direct your focus on gathering evidence about what the students seem to understand about surface area. With your groupmates, develop at least three statements related to students' understanding. Your statements should communicate something specific about what you think a particular student or small group of students understands. You should give evidence, on the basis of students' written work or spoken comments in the videos, to support your claims.

(4) During one of the later video clips, the teacher stated, "At this point in the lesson I felt pretty comfortable that most of the children had a pretty good understanding that there were ways that were more efficient than others to find the surface area of rectangular prisms, and so at that point I wanted to go after one more big idea from this lesson." What do you think contributed to the teacher's sense that the students had a good understanding about finding surface area? What particular strategies did he use to investigate the students' understanding during the lesson?

(5) What lingering questions about students' understanding do you have after watching the video clips? If you were the teacher in this classroom, what strategies might you use to further assess the nature of your students' understandings?

Activity 6.2
Examining Assessment Resources in Curriculum Materials

In this activity, you will examine the assessment resources from two different curriculum programs. NCTM (1995, p. 3) defines assessment as "the process of gathering evidence about a student's knowledge of, ability to use, and disposition toward mathematics and of making inferences from that evidence for a variety of purposes." As you review curriculum materials, you will identify resources and techniques that support teachers in gathering such evidence.

(1) Obtain the teacher's materials for a unit from a Standards-based curriculum program. Briefly describe the unit you have chosen. Include the name of the curriculum program, the unit or chapter, the mathematical goals or emphases, and grade level.

(2) List the assessment resources in the unit or in any accompanying teacher materials (e.g., an assessment handbook). Be sure to include assessment resources in addition to tests and quizzes. As *Principles and Standards* (NCTM 2000, p. 23) explains, "In addition to formal assessments, such as tests and quizzes, teachers should be continually gathering information about their students' progress through informal means, such as asking questions during the course of a lesson, conducting interviews with individual students, and giving writing prompts."

(3) (a) Which assessment resources and techniques from the unit would help you *during* instruction? How?

(b) Choose one informal assessment resource or technique from part (3)(a) and describe *specifically* what information about students' understanding you could obtain by using it during instruction.

(4) (a) Which assessment resources and techniques from the unit would help you *after* instruction? How?

(b) Choose one assessment resource or technique from part (4)(a) and describe *specifically* what information about students' understanding you could obtain by using it after instruction.

(5) Select another unit from a different curriculum program. This time, you may choose to examine another Standards-based curriculum unit, from a different program, or a chapter from a commercially developed textbook.

(a) Describe the unit or chapter, as in question 1.

(b) List the assessment resources provided in the unit or in any accompanying teacher's materials (e.g., an assessment handbook).

(c) How do the assessment resources from this curriculum program or textbook series compare with those of the curriculum program you examined in questions 1–4?

(6) (a) NCTM (2000, p. 22) has suggested that "the tasks used in an assessment can convey a message to students about what kinds of mathematical knowledge and performance are valued." For both the units or chapters that you examined, state what kinds of knowledge and performance seem to be valued.

(b) NCTM (2000, p. 22) has also suggested, "When teachers use assessment techniques such as observations, conversations, and interviews with students, or interactive journals, students are likely to learn through the process of articulating their ideas and answering the teacher's questions." Of the two units or chapters you examined, which one recommended assessment techniques that would offer more opportunities for students to articulate their understandings?

CHAPTER 7

The Technology Principle: Using Standards-Based Curriculum Materials to Learn to Teach Mathematics with Technology

Susan M. Hagen

MOST preservice teachers (PTs) use technologies such as Facebook, MySpace, Twitter, texting, instant messaging, Xbox LIVE, Gmail, iPods, and cell phones daily. Although PTs have typically used a graphing calculator and perhaps Excel in previous mathematics courses, few are familiar with the wide range of technologies currently available for the mathematics classroom—the Geometer's Sketchpad, the WolframAlpha website, the TI-NSpire handheld calculators, the TI-Navigator system, accelerometer-based applets on the iPad, interactive whiteboards, Google documents, Microsoft Web applets, blogs, and wikis, to name just a few. As mathematics teacher educators, we need to consider how to familiarize PTs with an ever-changing array of available technologies and help them capitalize on the potential of these technologies for mathematics teaching and learning.

PTs need to engage with technology in ways that allow them to develop more than a superficial familiarity with it. Ideally, teacher education programs have an opportunity to help PTs understand the potential role of technology in contributing to students' learning. As Hollenbeck, Wray, and Fey (2010, p. 275) describe,

> Teachers need to carefully select and design learning opportunities for students where technology is an essential component in developing students' understanding, not where it is simply an appealing alternative to traditional instructional routines. To maximize the power of technology, teachers need access to resources and professional development opportunities to acquire a well-developed knowledge base for teaching with technology.

Here I share examples of how I have engaged the PTs enrolled in my mathematics course, Secondary Mathematics with Technology, with strategies and resources for effectively using new technologies in mathematics teaching and learning. In particular, I describe three course activities that have PTs use Standards-based curriculum materials as a context for learning about technology's role in mathematics instruction.

The Technology Principle

According to the Technology Principle (National Council of Teachers of Mathematics [NCTM] 2000, p. 25), "Technology is essential in teaching and learning mathematics; it influences the mathematics that is taught and enhances students' learning." An article in *Mathematics Teaching in the Middle School* (Edwards and Ozgun-Koca 2010) illustrated this Principle powerfully. The authors considered technology's role in an activity in which students examine how changing the parameters for the function $f(x) = ax^2 + bx + c$ affects the graph. The authors contrast what this examination involved in 1959, when students graphed by hand on paper, with how spreadsheet and graphing-calculator technology have affected the activity (p. 460):

> Even using circa 1989 graphing calculators, the activity is extended by considering two effects simultaneously. It is enhanced by the ease with which the calculator draws the graphs. Using technology clearly makes the time-consuming, tedious process of graphing more feasible.

Introducing second-generation graphing calculators such as the TI-NSpire has enhanced students' learning further. With the TI-NSpire, students can observe directly how changing the algebraic form changes the graph *and* vice versa.

As this example illustrates, technology makes some mathematics more accessible and, at the same time, makes some mathematics less important in the school curriculum (NCTM 1989, 2000). An important goal of my technology course is to help PTs understand how technological developments affect students' potential for learning in mathematics. My students and I have considered the following questions together:

- How does technology influence what mathematics we should teach?
- How does technology otherwise affect mathematics teaching?
- How does technology affect students' learning?
- How do technology-intensive mathematics classrooms differ from those in which we may have learned mathematics?

As PTs gain experience with technology in mathematics instruction, their responses to these questions typically become more sophisticated. For instance, many PTs come to my technology class expecting to teach in ways similar to how they were taught mathematics. They also bring genuine concerns about students who seem overly reliant on calculators for basic facts and procedures. Over the semester's course, we explore these views and concerns by questioning and revisiting PTs' initial assumptions, particularly about what mathematics is essential and what it means for students to learn mathematics.

In the three sections that follow, I discuss three activities from my Secondary Mathematics with Technology course, each of which uses Standards-based curriculum materials in its own way. The first project engages PTs in conducting curriculum reviews to identify the technology's role in different curriculum materials. In the second, I describe PTs' initial efforts to adapt existing mathematics lessons, drawn from textbook and curriculum materials, to incorporate technology meaningfully. The third project requires PTs to consider how to assess students' work that emerges from technology-rich mathematics lessons.

Computer and Calculator Software and Activities That Augment Curriculum Materials

Early in the semester in my Secondary Mathematics with Technology course, I have two primary goals. The first is to help the PTs feel comfortable with using different technologies, including graphing calculators, applets, motion detectors, handheld calculators, and spreadsheets. In other words, we have PTs play with a variety of technologies to gain familiarity with their capabilities (Hollebrands and Zbiek 2004). The second is to challenge PTs to begin thinking about where these technologies fit into the mathematics curriculum.

One way I have approached this second goal is to have PTs explore both lesson exchange websites and the technological tools that accompany curriculum programs. Almost any mathematical concept has corresponding technology-rich activities readily available online. Websites such as Shodor (http://www.shodor.com), the Texas Instruments Activity Exchange (http://education.ti.com), and NCTM's Illuminations (http://illuminations.nctm.org) are just a few sites where teachers can access high-quality lesson plans. Most Standards-based curriculum materials include activities that use technology to enhance students' learning and offer resources to supplement curricular tasks. For example, the Core-Plus Mathematics Project (CPMP) curriculum program offers a suite of Java-based applications that support the program's algebra, geometry, statistics, and discrete mathematics strands. One can download these tools from the Internet through the companion website, http://www.wmich.edu/cpmp/CPMP-Tools. The algebra tools include an electronic spreadsheet and a computer algebra system (CAS). Other textbooks, such as the Connected Mathematics Project, include applets available from their website, http://connectedmath.msu.edu/CD/index.html, or by purchasing them on a CD. These applets correspond to activities included in the curriculum materials. Working with these tools and applets, PTs can begin to identify ways that particular technologies may be powerful for supporting students' learning of particular mathematical concepts or procedures.

Later in the semester, PTs complete a curriculum review, similar to the curriculum reviews from this volume's previous chapters. The PTs must choose one of four particular mathematics topics (linear functions, quadratic functions, solving systems of equations, and proportional reasoning) from the algebra strand of the secondary curriculum (CCSSI 2010; NCTM 2000). They must also choose how to introduce the topic to students in three independent curriculum programs, one commercially developed and two Standards based. The topics chosen for PTs to analyze are ones that we have already investigated with technology in class. The topics are broad enough to show a progression of lessons but narrow enough to let the student focus on the methods' specifics.

The review's main focus is to discuss technology's use, or lack thereof, in a sequence of three to four consecutive lessons in each curriculum program. Working in small groups, PTs select sequences of lessons from each curriculum program and respond to questions such as the following:

- What mathematical topics does each lesson sequence address? How does each one address them?
- What are the goals or objectives for students?
- What is the lesson activities' progression?
- Do the activities use technology? What kind? How? Do the activities use the technology effectively?
- How might you modify the lesson to incorporate technology more effectively?
- How are the lessons in the three programs similar? How are they different?

I remind the PTs to include specific examples to illustrate their claims, as well as to review both the students' and teacher's lesson pages. (In this project's first year, PTs found little use of technology in the materials. I realized that if the students' and the teacher's editions were not combined into one book, as with the Connected Mathematics Project series, the PTs had trouble connecting the two. When the suggestions for technology use were only in the teacher's edition, the PTs saw them as optional. The project now stresses looking at both sources as well as at the online supplements.) I have used this activity toward the end of the course for several years, and each year the projects have improved.

With technology ever changing, curriculum designers find integrating specific technologies into students' editions difficult. As suggested earlier, PTs commonly indicate that the suggestions for technology in the teacher's manual were "add on" and not integrated into the concept's development. For example, in their critique on solving systems of equations, one group of PTs wrote the following:

> Technology use is given as a supplement to many lessons but is not included as the fundamental applet for instruction. For example, one lesson has the option for students to use the calculators to graph the equations and see where they intersect. Students learn to graph the equations and find the intersection without technology and they discuss what the intersection means. However, this discussion isn't mentioned when technology

is listed as an option. They simply have students graph the equations with the calculator but nothing more. . . . It seems they simply use the calculator to support graphing the systems of equations by hand.

Although we had worked with a Calculator-Based Laboratory (CBL) and motion detectors in class, the PTs seemed to want more direction and guidance than the textbook offered. They were interested in details about how to operate the technology and what to expect from students in class. Similarly, another group made the following remarks:

> The curriculum glosses over technology with a suggestion to use CBL, but with no ideas as to how to use it. Graphing calculators are mentioned as being helpful for solving systems of linear equations but without specific suggestions for utilizing them in the classroom.

On the basis of these writings, it appears that the more guidance a textbook gives, the more likely the PTs are to view the technology as integral and be confident that the activity uses technology effectively. As one group of students wrote,

> This book gives the most in-depth explanation of technology use. Technology is intertwined throughout the whole lesson in asking students to do specific tasks with either the graphing calculator or the CAS. Screenshots of calculators are printed in the teacher's edition to give some guidance.

This written guidance also helps confirm for PTs that using technology in this way is an essential part of learning mathematics.

Each year, as the PTs have worked through this activity, they have concluded that even the newer curriculum lacked the necessary guidance on technology use. Therefore, activities like this curriculum review are crucial. They must help PTs develop the inclination and knowledge to adapt written curriculum materials, so that PTs develop lessons that use technology as an integral part of the learning process, rather than as something to add on afterward.

Adaptations of Lessons from Curriculum Materials to Incorporate Technology

The Technology Principle demands not only that PTs develop an appreciation for the power of computers, calculators, and other technological tools but also that they learn to acquire the knowledge and confidence needed to select and use technology effectively in instruction and assessment. Although the available resources are tremendous, PTs in my classes communicate the view that technology-rich lessons are most appropriate *after* students have been introduced to a topic traditionally. This view seems to stem, at least in part, from how the PTs themselves learned mathematics. To change this perception, we need to convince PTs that, often, using technology can be more powerful, efficient, and engaging than familiar pedagogical approaches. By creating opportunities for PTs to connect available activities that include technology with effective mathematics curriculum materials, we can have them begin to distinguish between familiar teaching strategies and the variety of ways that teachers can use technology to make learning mathematics more meaningful for students.

The secondary school mathematics teacher education program at Virginia Tech includes a two-course series, Mathematics for Secondary Teachers, in which PTs explore secondary school mathematics curriculum in part through using middle and high school Standards-based curriculum materials. Usually approximately half the students in my technology course are also enrolled in one of the Mathematics for Secondary Teachers courses. For several years, I collaborated with these courses' professor, Gwendolyn M. Lloyd, to support a course project with a technology component. This collaboration began because, initially, Gwen assigned a project that required PTs to incorporate technology into curriculum-based lesson plans that they developed for her course. As the technology course's instructor, I received many e-mails from PTs asking whether I knew of an applet or computer program that could satisfy this requirement. Or, they stopped by and spent a few minutes talking with me about ideas. Then, they invariably used the technology only superficially in their lessons. With a few exceptions, PTs incorporated technology into their lessons because they had to, not because they thought it was essential to learning.

To help the PTs incorporate technology into lessons meaningfully, Gwen and I developed a new approach. As shown in activity 7.2, Gwen required the PTs in her class to form groups such that at least one PT in every group was also enrolled in my technology class. The PTs had to present a preliminary version of the lesson in my technology class, before presenting the lesson in Gwen's Mathematics for Secondary Teachers class. PTs modified their lessons' technology component on the basis of feedback from my class. In their written reflections after presenting the lesson, they wrote about that feedback and how they adjusted the lesson in accordance with it.

This collaboration gave PTs opportunities to develop effective ways of using technology in their lessons. It also improved the overall quality in the PTs' lessons of both the mathematics and the pedagogy. Consider one PT's presentation to my class about a lesson on the sum of a polygon's interior angles. The PT set up a presentation using hide/show buttons in Geometer's Sketchpad to demonstrate how a particular figure could be cut into triangles to derive the formula $(n - 2) \times 180°$. His figures were "precut" with a hide/show button and progressed so that he led them to "see" that an n-sided polygon could be cut into $n - 2$ triangles, the angles of each of which summed to 180 degrees. Figure 7.1 shows one shape with six sides and four triangles. He dragged the figures to show that the sum of the angles was always the same.

The Technology Principle

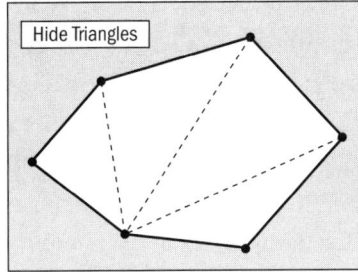

Fig. 7.1. A shape with six sides and four triangles

One student in the class asked what would happen if the triangles were drawn differently and drew a picture like the one in figure 7.2 on the board.

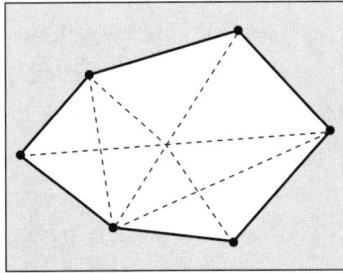

Fig. 7.2. A student's revision of the triangles shown in figure 7.1

This student saw the hexagon as made up of six triangles, rather than the original four triangles. Initially, the presenter's reaction was that he could not use this picture. After much discussion, however, he realized that this method also worked and related to the formula he had from the previous set of figures by the distributive law. He realized that the new picture represented $n \times 180° - 360°$. This experience in the technology course offered the PT the opportunity to think further about his group's lesson ideas and to guide his group in making necessary revisions to their project.

Assessment with Technology

Learning to teach using technology requires that PTs not only learn how to use tools to develop students' understandings but also develop strategies for assessing those understandings. Using an interactive geometry program permits several approaches to the same problem. Lessons in many Standards-based curriculum programs encourage students to "use a geometry drawing program" or "dynamic software" to investigate properties of geometric figures.

Activity 7.2 was designed to introduce the TI-NSpire handheld "Graphs and Geometry" program by determining how one can use the tools differently to construct sketches that look the same on the surface but behave differently when dragged. Although this activity was written for the TI-NSpire handheld, it can easily be modified for any interactive geometry program. (To complete activity 7.2, PTs will need a .tns file loaded into their TI-NSpire or TI-NSpire CAS. The EqualConstructions.tns file is available for download at the TI Activities Exchange website, http://education.ti.com/educationportal/activityexchange/Activity.do?cid=US&aId=13208.)

This activity introduces PTs to the circle and line tools (i.e., ruler and compass) in the TI-NSpire geometry application with an eye toward developing a way of assessing geometric constructions. They begin by looking at different ways to construct intersecting circles with sketches that have been drawn for them. First, we tell them how we drew the sketches. They then explore the control points for size and for movement without a change in size. Later, we reverse the process and have them determine how to construct a sketch on the basis of where they determined the control points were and how they moved the figure without changing the size. Finally, they look at sketches of tangent lines to determine what the sketch can tell them about what a student knows about tangents. A modified version of the problem, from Scher's (2005) chapter, "Square or Not," duplicates the process on squares. The PTs conclude the activity by reading Scher's chapter.

Because of activity 7.2's richness, I have found many things to discuss and consider after PTs' work on it. My PTs and I have discussed difficulties in determining how to construct the sketches, as well as similarities and differences among the sketches. An important question has been, "what determines a sketch's correctness?" Sometimes, that one sketch is more correct than another is not clear. One can also extend activity 7.2 to include creating rubrics for evaluating dynamic sketches. I have had PTs develop rubrics for the tangent line problem in activity 7.2. First, however, PTs classify the tangent line sketches according to Scher's (2005) classification system. After classifying the sketches and developing a rubric for evaluating them, PTs revisit—and revise—their original analysis of students' understanding of tangent lines.

I also developed the following project, related to Standards-based and commercially developed curriculum materials, as another follow-up assignment to activity 7.2:

> Choose one topic (tangent lines, similar polygons, or properties of a square). Identify how your topic is introduced in two different curriculum programs, one commercially developed and one Standards based. If the lessons do not include the use of technology, describe *in depth* how you can modify the lesson to include an interactive geometry component. If the lessons do include technology, what modifications would you make based on your work with the TI-NSpire activity?

But most important, after working through the activity, PTs must investigate how secondary school curriculum teaches tangent lines, squares, and similar figures, as well as how, with this activity, they can modify that approach to enhance students' understandings.

Conclusions

Embracing the Technology Principle requires that PTs be critical, creative consumers of many technological and instructional resources and that they apply their mathematical and pedagogical knowledge to develop and adapt lessons to incorporate technology into mathematics instruction. Today's PTs and their future students will have access to technologies and curriculum resources yet to be designed. Because teachers may find keeping up with rapid changes in specific tools and materials dificult, the activities that I described here do not specify one technology or curriculum program. Instead, they emphasize the importance of learning both how to use technologies and how technologies can enhance students' learning of important mathematics.

References

Common Core State Standards Initiative (CCSSI). *Common Core State Standards for Mathematics.* Washington, D.C.: National Governors Association Center for Best Practices and the Council of Chief State School Officers, 2010. http://www.corestandards.org.

Edwards, Thomas G., and S. Asli Ozgun-Koca. "A Historical Perspective from A to C." *Mathematics Teaching in the Middle School* 15 (April 2010): 458–65.

Hollebrands, Karen, and Rose Mary Zbiek. "Teaching Mathematics with Technology: An Evidence-Based Road Map for the Journey." In *Perspectives on the Teaching of Mathematics,* 2004 Yearbook of the National Council of Teachers of Mathematics (NCTM), edited by Rheta N. Rubenstein, pp. 259–70. Reston, Va.: NCTM, 2004.

Hollenbeck, Richard M., Jonathan A. Wray, and James T. Fey. "Technology and the Teaching of Mathematics." In *Mathematics Curriculum: Issues, Trends, and Future Directions,* 2010 Yearbook of the National Council of Teachers of Mathematics (NCTM), edited by Barbara J. Reys and Robert E. Reys, pp. 265–76. Reston, Va.: NCTM, 2010.

National Council of Teachers of Mathematics (NCTM). *Curriculum and Evaluation Standards for School Mathematics.* Reston, Va.: NCTM, 1989.

———. *Principles and Standards for School Mathematics.* Reston, Va.: NCTM, 2000.

Scher, Daniel. "Square or Not? Assessing Constructions in an Interactive Geometry Software Environment." In *Technology-Supported Mathematics Learning Environments,* 2005 Yearbook of the National Council of Teachers of Mathematics (NCTM), edited by William J. Masalski, pp. 113–24. Reston, Va.: NCTM, 2005.

Activity 7.1
Mathematics Teaching Project

As the chapter text described, Gwen Lloyd's Mathematics for Secondary Teachers course used this activity, which I share here with permission.

This semester you will complete a teaching project, from your choice of one of the three areas of the course (geometry and trigonometry, statistics, and probability). Each teaching project consists of a written component and a presentation. A teaching project involves using a middle or high school textbook to prepare and teach a lesson related to one of the mathematical areas we are studying.

Groups

- You will work in a small group of two to three people on your teaching project.

- You will submit just one group copy of the project report. However, each member of the group must submit, individually, a copy of a "Reflection on Group Work."

- Because your lesson must include technology (see below) and your project requires feedback from the Secondary Mathematics with Technology class, you should form groups so that at least one group member is currently enrolled in Secondary Mathematics with Technology.

Materials

The first step in a teaching project is choosing a lesson to focus on. Each area of the course (geometry and trigonometry, statistics, and probability) has a list of textbook lessons from which you can choose. The list includes middle school and high school textbook lessons. You will need to choose one lesson or activity that interests you and your group.

The lesson lists and the actual textbooks are kept in our classroom, so you should allocate some time before or after class to look over several of these lessons as you make your choice.

Presentation/Teaching

Once you have chosen a lesson or part of one from the list, you can begin preparing to teach the lesson to our class. The next section, which describes the written report you must submit, lists many of the steps involved with preparing to teach.

When you teach our class, you should plan to use 30–35 minutes of class time. Your presentation should consist of each of these elements, not necessarily in this order:

1. Present information about the source of the lesson you chose (the textbook), the lesson's level and topic, and how the textbook lesson relates to NCTM's (2000) Content and Process Standards for this area and level. (Note: This component of the project could be adapted to require PTs to compare the textbook lesson to the recommendations of the *Common Core State Standards* [CCSSI 2010] as well as the NCTM [2000] *Principles and Standards*.)

2. Explain *how* and *why* you changed or adapted the textbook lesson. For example, did you add technology to the lesson, include more group work, or change the mathematical emphasis of the lesson?

3. Actually teach your lesson, or an important part of it, to the class. Your lesson should engage the class as if the students were in middle or high school, depending on your lesson's level. The lesson *must* include using either graphing calculators or computers and involve both small-group and whole-class components. You must base the lesson on a textbook lesson from the list, but you can adapt it to suit your interests, your goals, and this project's requirements.

Written Report

The written report should be typed and should include the following elements, preferably in this order:

1. Present information about your lesson's source (the textbook), its level and topic, and how the textbook lesson relates to NCTM's (2000) Content Standards and Process Standards for this area and level.

2. Explain *how* and *why* you changed or adapted the textbook lesson to teach our class. For example, did you add technology to the lesson, include more group work, or change the mathematical emphasis of the lesson? What were the goals or advantages of your changes? Indicate whether you would make these same adaptations if you had been teaching a middle or high school class. What might be different or the same?

3. Reflect on technology. At least one of your group members will receive feedback from the Technology class about a preliminary version of your group's lesson plan. What feedback did your group receive about your preliminary plan? What changes did you make in response to this feedback?

4. Include information about your actual lesson plan's objectives, materials, and technology and about the lesson activities' progression. Describe the students' and teachers' activities enough so that someone else could carry out the lesson plan. If you used handouts in your lesson with the class, include those.

5. Reflect on teaching the lesson. How did things go when you taught the class? Did the students engage with your lesson in the ways you thought they would? With what parts of the lesson were you happy or disappointed? What would you do differently if you were to teach this lesson in a real middle school or high school classroom?

6. Give two problems or questions that you think could be solved or answered by students who participated fully in your lesson. For each problem, include the problem and the solution, and say why this is an important problem that assesses your lesson's big ideas.

7. What do you believe you learned from doing this project? For example, did you learn any new mathematics? Did you learn anything about lesson planning, teaching, or textbooks?

8. Give the URL for a site where you have posted some part of this project online. Possible parts to post include parts 1–3, 4–5, and 6. We encourage you to post as much of your project as possible, in as organized a manner as possible, so that others can access it easily.

9. Reflect on group work. Each person in the group should write and submit a one-page report describing how you believe the project went, from start to finish. For example, what do you think about the lesson you taught? How did your group work together? What, if anything, do you wish had gone differently?

The Technology Principle

Activity 7.2
Are All Constructions Created Equal?

Using an interactive geometry program allows several approaches to problems. This activity is designed to familiarize you with the circle, compass, and line tools and how constructions may look the same but be different underneath.

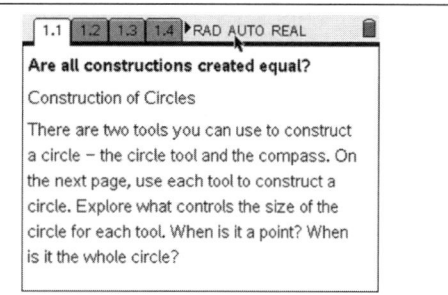

First, open the TI-NSpire document EqualConstructions.tns.

Then press Ctrl and right (on the trackpad) to move to page 1.2 to begin the lesson. See page 88 for more information on where to find their file.

1. Problem 1: Constructing circles

2. You can use two tools to construct a circle—the circle tool and the compass. Using the circle tool constructs a circle by setting the circle's center with the first click and its radius with the second. The compass tool constructs a circle from a center point, with a radius defined by a segment or distance measurement. On page 1.2, use each tool to construct a circle. You can access both tools through the Menu button. Explore what controls the circle's size for each tool. When is it a point? When is it the whole circle?

Write your findings in the table below.

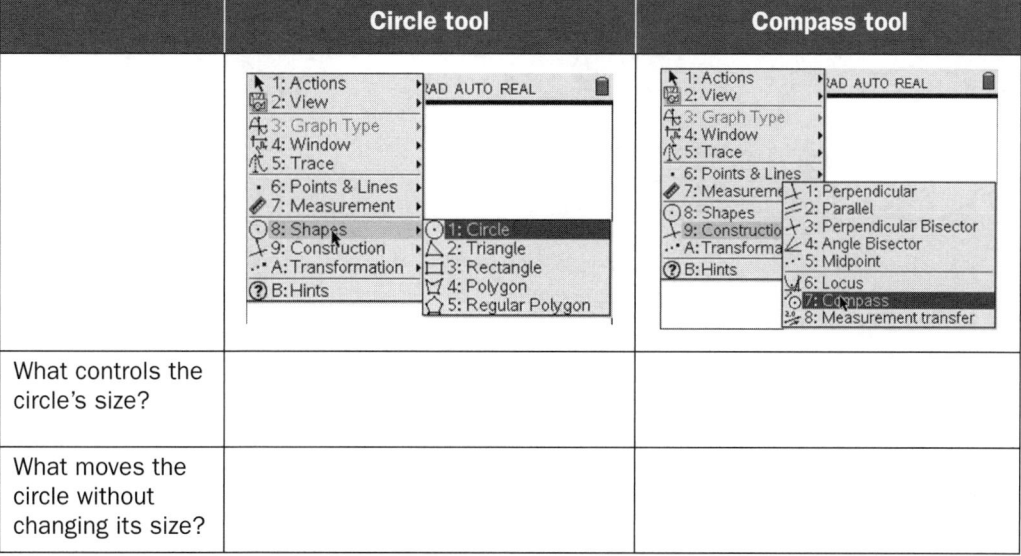

	Circle tool	Compass tool
What controls the circle's size?		
What moves the circle without changing its size?		

Curriculum-Based Activities and Resources for Preservice Math Teachers

3. Many ways exist for constructing two intersecting circles. Pages 1.4–1.8 show examples of two intersecting circles that look the same but are constructed differently. Note how each figure is different and what controls the size of the circles. Write your observations in the table on the following page.

			What controls the circles' size?	What moves a circle without changing its size?
1		The two circles were constructed by using the Circle tool. Both circles have radius equal to $\overline{AB}$. Circle A was constructed first.		
2		The two circles were constructed by using the Compass tool. Point A was constructed first, point B was added, and the distance between point A and B was used for both circles' radii.		
3		Point A was constructed first by using the Point tool. With the Circle tool, point B was constructed and point A clicked.		
4		A segment $\overline{AB}$ was constructed first (hidden). The Circle tool was then used to construct two circles with centers A and B and radius $\overline{AB}$.		
5		A segment $\overline{A'B'}$ was constructed first. The Compass tool was then used to construct two circles with centers A and B and radius $\overline{A'B'}$.		

94

4. Problem 2: Shapes that maintain their characteristics

    Problem 2 investigates constructing triangles. Page 2.2 shows a triangle that maintains its shape no matter how you move the points (i.e., all the triangles that you construct as you move the points are similar to one another). This construction is based on the intersection of two circles. The circles are hidden, but you can see them by using the Hide/Show tool. Choose Menu, Actions, Hide/Show. The circles should be visible but dimmed in the background. On the basis of your investigation above, how could these circles have been constructed? What kind of triangle is $\triangle ABC$? Justify your choice.

    _____

    _____

    _____

    _____

    _____

    _____

Our constructions above were based on the circles. We can construct other kinds of triangles that remain similar, but we'll need more tools.

5. Constructions with lines: Use page 2.4 to experiment with the line and segment tools in the Points & Line menu. Notice how the line changes when you grab the line or the point. What is the difference between a line and a segment? Write your findings below.

    _____

    _____

    _____

    _____

    _____

6. On pages 2.5 and 2.6, you will see △ADE, which you can manipulate to form similar triangles. Notice that in both instances, point D is the control point for size change. What do point A and E control for each? These constructions are based on lines and circles. Again, to see the construction's hidden parts, use the Hide/Show tool. The circle here depends on the ratio of the segments' measurements to guarantee similarity. △ABC forms the original triangle. Show the points A and C, the sides of the triangle, and the lines AB and AC by clicking them. When the points, sides, and lines are no longer dim, press Escape. Your picture should look similar to those in the table below. Let us focus on how the lines were constructed and used. In the table below, read how the sketches were constructed and note the differences in how the triangles can be moved or reshaped by using the points, the lines, or the segments.

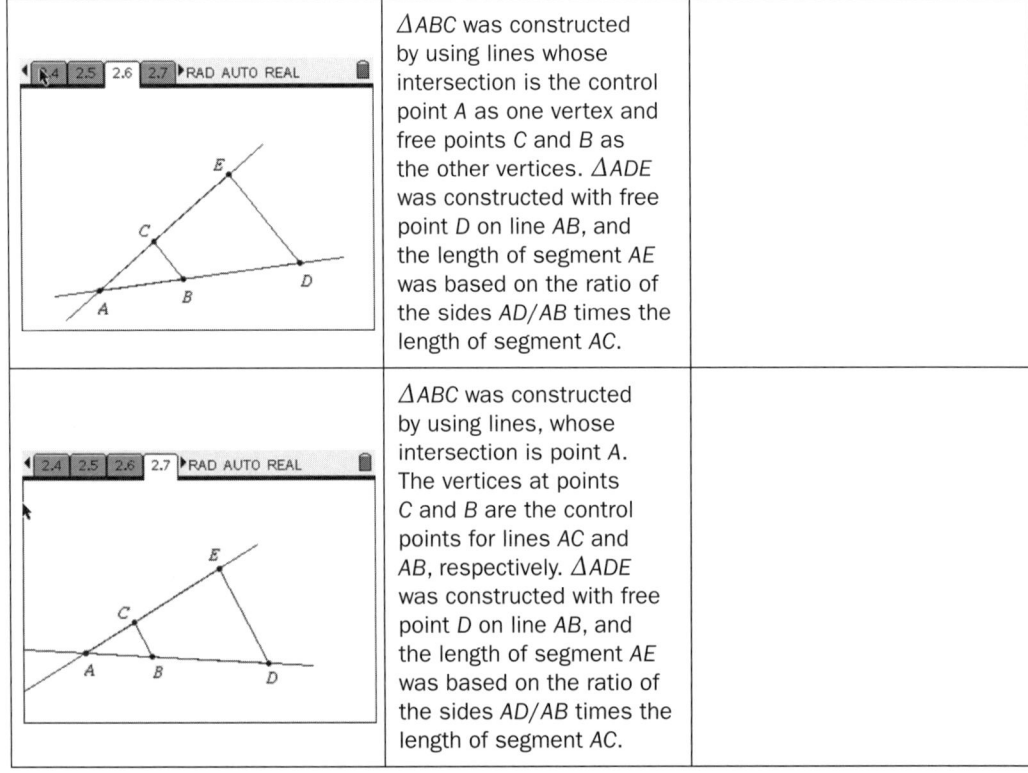

	△ABC was constructed by using lines whose intersection is the control point A as one vertex and free points C and B as the other vertices. △ADE was constructed with free point D on line AB, and the length of segment AE was based on the ratio of the sides AD/AB times the length of segment AC.	
	△ABC was constructed by using lines, whose intersection is point A. The vertices at points C and B are the control points for lines AC and AB, respectively. △ADE was constructed with free point D on line AB, and the length of segment AE was based on the ratio of the sides AD/AB times the length of segment AC.	

Describe how the sketches are similar.

Describe how the sketches are different.

7. Problem 3: Constructing tangent lines

Problem 3 looks at constructing tangent lines by four different methods. Determine how the sketches were constructed and what these constructions can tell you about the student's understanding of tangent lines. Write your observations in the table below.

	How was the sketch constructed?	What does the construction tell you about the student's understanding of tangent lines?

Curriculum-Based Activities and Resources for Preservice Math Teachers

8. Problem 4: Constructing squares

In problem 4, you will find four sketches of square *ABCD* that look the same but are constructed differently. The sketches were adapted from Scher (2005) and modified for the TI-NSpire.

Before reading the article, go through the four sketches and determine how they were constructed and what the construction can tell you about a student's understanding of a square. Write your observations below. Read the article and reflect on your observations.

	How was the sketch constructed?	What does the construction tell you about the student's understanding of the properties of squares?
[sketch 1: square ABCD]		
[sketch 2: square ABCD]		
[sketch 3: square ABCD]		
[sketch 4: square ABCD]		

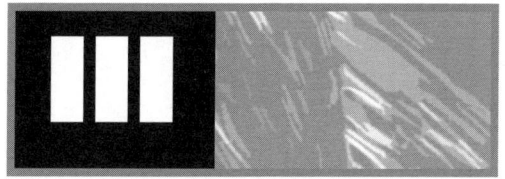

From the University to the Classroom

CHAPTER 8

Preservice Teachers' Learning in Grades K–12 Classrooms: Engaging with Standards-Based Mathematics Curriculum Materials in Field Experiences

Andrea V. McCloskey
Elizabeth Winarski

IN THIS VOLUME's preceding chapters, teacher educators offer insights about how one can use Standards-based curriculum materials to support the learning of preservice teachers (PTs). In particular, the authors share strategies and activities that they have used in their teacher education courses to engage PTs with the six Principles for school mathematics (National Council of Teachers of Mathematics [NCTM] 2000). One can modify many of these activities for a variety of contexts beyond university-based methods courses with PTs, including professional development activities with in-service teachers. Here we extend the consideration of mathematics teacher education beyond university coursework to another part of the teacher-learning trajectory: field experiences. We discuss what Standards-based curriculum materials can contribute to this important component of teacher preparation.

Field Experiences in Mathematics Teacher Education

Field experiences are a typical component of teacher education programs. PTs often complete a series of field experiences on their way to gaining certification or licensure to teach. PTs often point to their field experiences, especially their time as student teachers, as the most educative component of their teacher preparation programs (Feiman-Nemser 1983; Guyton and McIntyre 1990). One of our students in a mathematics methods course recently reported that she "learned more in one day of being in a classroom than in all [her] methods courses combined." Our first reaction to such statements is to wince and then gently suggest to the student that perhaps she could learn so much from being in a classroom because her university courses had prepared her to be able to learn. Nevertheless, we also think that university-based mathematics teacher educators must recognize that mentor teachers, also called *host teachers* or *cooperating teachers*, are our partners in preparing teachers. Our conversations about and around curriculum materials must acknowledge and explicitly draw on PTs' experiences in classrooms with students and mentor teachers.

Standards-Based Curriculum Materials' Role in One Field-Experience Classroom

Several studies of student teachers' use of mathematics curriculum materials (e.g., Lloyd 2008; van Zoest and Bohl 2002) have examined PTs as they use Standards-based materials in their classroom internships. Findings from these studies suggest that the relationship between the student teacher and the mentor teacher influences how the student teacher comes to use and regard curriculum materials as resources for mathematics instruction. One of us, Elizabeth Winarski, has served as a mentor teacher, hosting PTs who were completing early field experiences or student-teaching internships in her multi-aged kindergarten and first-grade classroom. In the following, Elizabeth reflects on what using a Standards-based program contributes to her mentoring role.

A Mentor Teacher's Reflection

To address all the district and state standards that I must, I use a variety of resources for teaching mathematics. However, Investigations in Number, Data, and Space [TERC 2008] is the main curriculum program that I use. As a Standards-based curriculum program, Investigations offers many opportunities for my students to learn important, powerful mathematics. It also offers many opportunities for the PTs who come to my classroom to learn important, powerful lessons about teaching mathematics. Sometimes the PTs are completing an early field experience, in which they come once a week in groups of four to observe and later lead a mathematics lesson with a small group of my students. Sometimes the PTs are completing a student-teaching internship, during which I work closely with them alone as their mentor teacher over several months. During the student-teaching internship, they gradually assume more responsibility in the classroom, so that eventually they plan and implement whole-class lessons. In either instance, I am convinced that the PTs benefit from my using a Standards-based curriculum program as my primary resource.

An immediate benefit for the PTs is that the students in my classroom are doing mathematics. They are not simply watching and following as their teacher solves mathematics problems, as they might in a classroom that strictly follows a commercially developed curriculum program. Instead, my students are solving authentic mathematics problems. They are posing questions, making and testing predictions, and evaluating their own and one another's reasoning. In keeping with the Learning Principle (NCTM 2000), they are learning "mathematics with understanding, actively building new knowledge from experience and prior knowledge" (p. 20). As a result, my conversations with the PTs whom I mentor most often focus on the mathematics that the students can do and what mathematical understandings we should help them develop next. In other words, mathematics plays a central role in my conversations with the PTs in my classroom, and I think I can attribute this to the central role that big mathematical ideas—both concepts and procedures—play in Investigations.

Furthermore, the PTs who come into my classroom have an immediate role to play. Because we work primarily from a Standards-based curriculum program, the students' mathematical activity is the central event in every mathematics lesson. This means that even if the PTs are completing an early field experience, have not yet completed a mathematics methods course, and are not yet ready to teach a mathematics lesson, they still have opportunities to learn about mathematics teaching meaningfully in my classroom. The students in my classroom have opportunities to engage in sustained problem solving, often in small groups. During this problem solving, I often have the PTs circulate, ask my students questions, and try to help my students without removing the mathematical challenge. One of the Standards for Mathematical Practice of the *Common Core State Standards* (CCSSI 2010)—make sense of problems and persevere in solving them—characterizes the mathematical activity of the children in my classroom. My students are used to explaining their reasoning; this is a norm that we have established. They are thus willing, and often eager, to talk with PTs about the mathematics they are doing. This approach strikes me as quite different from other types of field experiences in which PTs, while visiting classrooms before their student-teaching assignment, simply observe a lesson being delivered, because the lesson offered no opportunity for another teacher to contribute or participate meaningfully. In those more traditional sorts of field experiences, the curriculum simply dictated what mathematics the teacher should present, and the students would then practice the procedures that had been taught on their own. PTs were left to observe. Although things certainly occurred that the PTs could notice and learn from, they were not necessarily observing compelling mathematical activity.

I also value using a Standards-based curriculum program because it helps me model the Teaching Principle (NCTM 2000) to the PTs in my classroom. The Teaching Principle characterizes effective teaching as requiring "continuous efforts to learn and improve . . . about mathematics and pedagogy" (p. 19). Standards-based curriculum materials, by design, require and support teachers' learning. One cannot master these materials simply by teaching them once. I learn continually through the materials and the children's mathematics, and I appreciate that I can model this orientation toward lifelong learning for PTs. Another aspect of the Teaching Principle is that "there is no one 'right way' to teach" (p. 18). The PTs who come into my classroom observe me using a variety of teaching styles and strategies. Sometimes they see me play a large role in a whole-class presentation—for example, when I introduce a new mathematical term. Other times, my students work independently in groups, and my role is less central. Still other times, I talk one on one with a student to assess what he or she understands. PTs can thus observe that effective mathematics teaching involves a variety of teacher–student interaction patterns.

In my experience, classrooms that use Standards-based curriculum programs as a primary resource tend to treat NCTM's (2000) *Principles and Standards for School Mathematics* as integral to the everyday activity of learning mathematics rather than as items included so they can be "checked off." The students in my classroom communicate, represent, make connections, solve problems, and reason every day; we do these activities as we learn mathematics. I constantly attend to issues of equity, curriculum, teaching, learning, technology, and assessment. These Principles are not afterthoughts: in my classroom, we inextricably link them with the mathematics we study. The PTs who visit and participate in our classroom mathematics community engage with the full spirit of *Principles and Standards*, which using a Standards-based curriculum program makes more accessible to both them and me.

Revisiting Curriculum Activities during Field Experiences

Above, Elizabeth reflected on how using a Standards-based curriculum program supports her ability to mentor effectively the PTs who visit her classroom during their field experiences. In this section, we propose how we might present and extend two activities from earlier chapters during field experiences. We agree with Drake and Land when they write in chapter 4 that it is important "to view these activities not only as stand-alone tasks for use at any point in a methods course but also as part of a trajectory of PTs' and teachers' learning about curriculum materials." We offer a few suggestions of how we could revisit an activity from a methods classroom during a field experience.

Examples come from activity 3.2 and a similar activity from chapter 4, in both of which the PTs examine a set of curriculum materials and analyze tasks by using the Cognitive Demand of Tasks framework (e.g., Smith and Stein 1998). Herbel-Eisenmann's methods course introduces PTs to the framework and supports them as they learn to use this framework to classify the cognitive demand of tasks printed in written curriculum materials. Herbel-Eisenmann also mentions that she discusses with her PTs how they "might modify textbook problems to make the problems high-level." One way to extend this assignment into a field experience is to introduce the PTs to another part of the framework, which examines task phases. Smith and Stein explain that tasks pass through three phases: curriculum materials present them, teachers set them up, and students finally implement them. Classroom observations have revealed that "the nature of tasks often changes as they pass from one phase to another" (Smith and Stein 1998, p. 270).

Often, tasks written at a high level of cognitive demand are implemented in a way that removes, or at least somehow lessens, the mathematical challenge, and students do not engage in powerful reasoning. This decline of cognitive demand can happen for a variety of reasons, but characteristics and principles exist that can help teachers who work to maintain high levels of cognitive demand when setting up and implementing tasks.

We think that PTs who have been introduced to, and used, the Cognitive Demand framework to analyze written tasks (e.g., in a methods course) should analyze all phases of a mathematical task. For example, before observing how a mentor teacher teaches, the PT could analyze the task's cognitive demand as the curriculum materials present it. The PT might then watch closely as the teacher sets up the task and then observe as the students take up and engage with it. The PT could answer whether the activity maintained or lessened the task's cognitive demand level. What factors may have contributed to a change in the task as it passed through phases? If the task's cognitive demand did decline, what might the teacher have done differently to prevent this? Revisiting activity 3.2 in this way can reinforce the Teaching Principle's statement that "worthwhile tasks alone are not sufficient for effective teaching" (NCTM 2000, p. 19). Finding good tasks, or making effective modifications to tasks, is only part of ensuring that students engage with high-level tasks.

Activity 3.1, in which the PTs wrote a letter to a hypothetical school principal recommending or rejecting a specific set of mathematics curriculum materials for adoption, is also worth revisiting during a field experience. The PTs, imagining that they are serving on their current placement's textbook adoption committee, might find revisiting their letter—and answering some additional questions—valuable toward the end of their field experience. You might have PTs respond to questions such as the following:

- Now that you have gotten to know a school culture a little better, how might you change your letter?
- Would you change your recommendation for adoption or rejection?
- Would you change your evidence or examples to make your letter more convincing?
- Would you recommend that the school adopt any support structures to make the curriculum adoption successful (i.e., changing the school day schedule to allow for teachers' planning time, purchasing new technological tools, or opening new lines of communication with parents)?

Reconsidering their textbook recommendation in light of a specific school placement could potentially increase PTs' appreciation for the value of using frameworks when evaluating curriculum materials.

The extensions presented above are just two examples of how the activities from earlier chapters can apply directly to field experiences. All the activities in previous chapters—and, indeed, most carefully designed experiences with Standards-based curriculum materials—can help lay a groundwork for developing curricular knowledge further while PTs complete field experiences. Simple exposure is not enough to encourage teachers to interact meaningfully with curriculum materials. Multiple exposures to Standards-based curriculum materials in increasingly authentic settings can help teachers learn to evaluate curriculum materials and ultimately use them as flexible resources. PTs' experiences with curriculum materials must be designed with the big ideas of mathematics teaching and learning in mind, as *Principles and Standards* (NCTM 2000) specifies. When PTs complete these experiences with thoughtful instructors', supervisors', and mentor teachers' guidance and support, PTs become able to use curriculum materials effectively. This scenario itself reflects the Curriculum Principle, with only slight adaptations to be relevant to mathematics teachers' preparation: to develop curricular knowledge for teaching mathematics, our teacher preparation programs must be coherent, focused on important principles of teaching and learning mathematics, and well articulated across university courses and field experiences.

Conclusions: Connecting Curriculum Activities across Mathematics Teacher Education

Along with this volume's other authors, we have called for mathematics teacher educators—and in this category, we include university-based course instructors, classroom-based mentor teachers, and supervisors—to consider ways that they can use Standards-based curriculum materials, both before and during field experiences, to maximize PTs' learning about and through textbooks. It will be interesting to see how increased exposure to Standards-based curriculum materials in teacher education affects how teachers engage with them in designing and enacting mathematics instruction over their careers. Considering the increasing number of PTs who encounter Standards-based curriculum materials, both as students in teacher preparation programs and as learners during their own grades K–12 mathematics education, we hope that researchers will explore this topic.

Two points to remember: First, even when activities such as those presented in the preceding chapters prepare PTs well, PTs will continue to face challenges when using Standards-based curriculum materials in their own classrooms. Second, PTs will also face these challenges even when a teacher like Elizabeth, who uses a Standards-based curriculum program carefully, has mentored them well. Standards-based curriculum materials' nature and design require teachers' ongoing learning. Teachers who use these materials to their full potential must continually learn more about mathematics and teaching mathematics (Ball and Cohen 1996; Lloyd 1999, 2002; Remillard 2000). For this reason, our support for teachers, as they continue to develop curricular knowledge, needs to extend beyond university courses, beyond field experiences, and into their teaching careers.

REFERENCES

Ball, Deborah L., and David K. Cohen. "Reform by the Book: What Is—or Might Be—the Role of Curriculum Materials in Teacher Learning and Instructional Reform?" *Educational Researcher* 25 (December 1996): 6–8, 14.

Common Core State Standards Initiative (CCSSI). *Common Core State Standards for Mathematics.* Washington, D.C.: National Governors Association Center for Best Practices and the Council of Chief State School Officers, 2010. http://www.corestandards.org.

Feiman-Nemser, Sharon. "Learning to Teach." In *Handbook of Teaching and Policy*, edited by Lee Shulman and Gary Sykes, pp. 150–70. New York: Longman, 1983.

Guyton, Edith, and D. John McIntyre. "Student Teaching and School Experiences." In *Handbook of Research on Teacher Education*, edited by W. Robert Houston, pp. 514–34. New York: Macmillan, 1990.

Lloyd, Gwendolyn M. "Two Teachers' Conceptions of a Reform-Oriented Curriculum: Implications for Mathematics Teacher Development." *Journal of Mathematics Teacher Education* 2 (October 1999): 227–52.

———. "Mathematics Teachers' Beliefs and Experiences with Innovative Curriculum Materials: The Role of Curriculum in Teacher Development." In *Beliefs: A Hidden Variable in Mathematics Education?*, edited by Gilah Leder, Erikki Pehkonen, and Guenter Törner, pp. 149–59. Utrecht, Netherlands: Kluwer Academic Publishers, 2002.

———. "Curriculum Use While Learning to Teach: One Student Teacher's Appropriation of Mathematics Curriculum Materials." *Journal for Research in Mathematics Education* 39 (January 2008): 63–94.

National Council of Teachers of Mathematics (NCTM). *Principles and Standards for School Mathematics.* Reston, Va.: NCTM, 2000.

Remillard, Janine T. "Can Curriculum Materials Support Teachers' Learning? Two Fourth-Grade Teachers' Use of a New Mathematics Text." *Elementary School Journal* 100 (March 2000): 331–50.

Smith, Margaret Schwan, and Mary Kay Stein. "Mathematical Tasks as a Framework for Reflection: From Research to Practice." *Mathematics Teaching in the Middle School* 3 (January 1998): 268–75.

TERC. Investigations in Number, Data, and Space. Glenview, Ill.: Pearson/Scott Foresman, 2008.

Van Zoest, Laura R., and Jeffrey V. Bohl. "The Role of Reform Curricular Materials in an Internship: The Case of Alice and Gregory." *Journal of Mathematics Teacher Education* 5 (September 2002): 265–88.

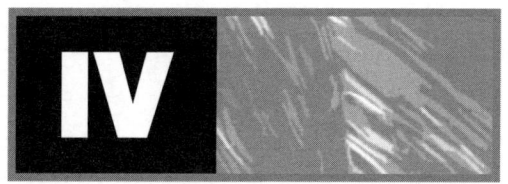

Resources

CHAPTER 9

Teacher Educators' Access to Curriculum Resources: History and Development of Standards-Based Programs, Sample Materials, and Technological Tools

Gwendolyn M. Lloyd

THIS CHAPTER'S purpose is to increase mathematics teacher educators' awareness of the range of curriculum resources that are available through the Internet. The chapter offers online resources in three areas: documents related to the history and development of Standards-based curriculum programs, sample materials from Standards-based curriculum programs, and technology. Although the particular websites in this chapter may change or vanish, we expect that publishers and curriculum developers will continue to offer more and more resources for teachers, parents, teacher educators, and curriculum decision makers in the coming years.

Resources Related to the History and Development of Standards-Based Curriculum Programs

To set a meaningful context for preservice teachers' (PTs') initial interactions with Standards-based curriculum materials, teacher educators often help PTs understand the history and philosophy behind these materials' development. *Curriculum and Evaluation Standards for School Mathematics* (National Council of Teachers of Mathematics [NCTM] 1989) and *Principles and Standards for School Mathematics* (NCTM 2000) have guided the development and revision of all Standards-based curriculum materials. These Standards documents, as well as other important curriculum resources, are available electronically on NCTM's website at http://www.nctm.org/standards/. This site also contains the historically important document *An Agenda for Action: Recommendations for School Mathematics of the 1980s* (NCTM 1980).

Another historical document available online is *Everybody Counts: A Report to the Nation on the Future of Mathematics Education* (Mathematical Sciences Education Board 1989, http://www.nap.edu/catalog.php?record_id=1199). For teacher educators interested in focusing on the curricular trends in mathematics education over time, the Center for the Study of Mathematics Curriculum, a National Science Foundation–funded Center for Teaching and Learning, has compiled a set of historical documents available for download and use in educational settings (http://www.mathcurriculumcenter.org/CCM/ccm_resources.php/). The recently developed *Common Core State Standards* (CCSSI 2010), which many states have now adopted, are available online (http://www.corestandards.org).

PTs may appreciate the opportunity to gain a broad view of the Standards-based curriculum programs currently available for the elementary, middle school, and high school grades. The K–12 Mathematics Curriculum Center (http://www2.edc.org/mcc/), funded by the National Science Foundation and housed at the Education Development Center, offers curriculum summaries, evaluation criteria, and strategies that teachers can use when selecting curriculum programs. PTs who plan to teach in high school may appreciate the website of Curricular Options in Mathematics Programs for All Secondary Students (COMPASS, http://www.ithaca.edu/compass/), which offers information about five high school mathematics programs. At the middle grades level, the Modeling Middle School Mathematics (MMM) project is a particularly valuable resource. The MMM video showcase (http://mmmproject.org/) offers classroom videos corresponding to lessons from five middle grades curriculum programs. In addition to the videos, the website offers students' and teachers' pages from the curriculum materials, as well as samples of students' work related to the video lessons.

Sample Materials from Standards-Based Mathematics Curriculum Programs

Engaging PTs in meaningful interactions with mathematics curriculum materials requires access to the curriculum materials themselves. Mathematics teacher educators have various levels of access to these materials: whereas some have access to an extensive curriculum library, others are limited to one or two copies of sample materials obtained from publishers. Regardless of one's personal access, the availability of sample materials and curriculum information *online* is an excellent resource for both teacher educators and their students. Figure 9.1 lists Standards-based curriculum materials, links to their websites, and additional links to available sample materials. Many publishers will supply samples of their materials for use in teacher education courses, even when not explicitly offered online.

Elementary Curriculum Program	Sample Materials
Investigations in Number, Data, and Space http://investigations.terc.edu/	http://investigations.terc.edu/curric-gl/
Everyday Mathematics http://everydaymath.uchicago.edu/	http://everydaymath.uchicago.edu/about/sample_lessons/

Curriculum Resources

Elementary Curriculum Program	Sample Materials
Math Trailblazers http://www.math.uic.edu/~imse/IMSE/MTB/mtb.html	http://www.kendallhunt.com/index.cfm?PID=2373&PGI=0
Middle Grades Curriculum Program	**Sample Materials**
Connected Mathematics Project http://connectedmath.msu.edu/	http://mmmproject.org/cmpS.htm (with classroom videos)
Mathematics in Context http://mathincontext.eb.com/	http://www.mmmproject.org/mic.htm (with classroom videos)
MathScape http://www2.edc.org/mathscape/	http://www.mmmproject.org/mathscape.htm (with classroom videos)
Math Thematics http://www.classzone.com/books/math_thematics1/	http://www.mmmproject.org/maththematics.htm (with classroom videos)
High School Curriculum Program	**Sample Materials**
Core-Plus Mathematics Project http://www.wmich.edu/cpmp/	http://www.wmich.edu/cpmp/course1.html http://www.wmich.edu/cpmp/course2.html http://www.wmich.edu/cpmp/course3.html http://www.wmich.edu/cpmp/course4.html
Interactive Mathematics Program http://www.mathimp.org/index.html	http://www.mathimp.org/curriculum/samples.html
Mathematics: Modeling Our World (ARISE) http://www.comap.com/highschool/projects/arise.html	http://www.comap.com/highschool/projects/mmow/C1sample.pdf
Math Connections http://www.its-about-time.com/htmls/mc/mcall.html	http://www.its-about-time.com/htmls/mc/mcbooksinbrief.html
SIMMS Integrated Mathematics: A Modeling Approach Using Technology http://www.simms-im.com/	http://www.simms-im.com/samples.html

Fig. 9.1. Sample materials from Standards-based mathematics curriculum programs

Technological Resources That Support Standards-Based Curriculum Materials

Because technology is a crucial component of effective mathematics instruction, it is also an important aspect of mathematics curriculum use (see chapter 7). In fact, many Standards-based curriculum programs explicitly incorporate technology and include technological resources for students and teachers. Figure 9.2 lists links to technology resources that accompany several Standards-based curriculum programs. Although one can explore these technological tools independently of the curriculum programs, they are most powerful when used to support intentional learning programs.

Elementary Curriculum Program	Technology Resource
Investigations in Number, Data, and Space	Information about the software used in several units: http://investigations.terc.edu/components/software/
Everyday Mathematics	Online games: http://media.emgames.com/emgames/demosite/demolevel2.html
Math Trailblazers	Sample online games: http://www.kendallhunt.com/index.cfm?PID=7951
Middle Grades Curriculum Program	**Technology Resource**
Connected Mathematics Project	Interactive Resources: http://connectedmath.msu.edu/CD/index.html
MathScape	Technology Options: http://www.glencoe.com/sec/math/mathscape/2005/course1/additional/index.php http://www.glencoe.com/sec/math/mathscape/2005/course2/additional/index.php http://www.glencoe.com/sec/math/mathscape/2005/course3/additional/index.php
Math Thematics	Online Tutorials and eWorkbooks: Book 1: http://www.classzone.com/books/math_thematics1/ Book 2: http://www.classzone.com/books/math_thematics2/ Book 3: http://www.classzone.com/books/math_thematics3/
High School Curriculum Program	**Technology Resource**
Core-Plus Mathematics Project	CPMP Tools: http://www.wmich.edu/cpmp/CPMP-Tools/

High School Curriculum Program	Technology Resource
Interactive Mathematics Program	Game of Pig Software: http://www.mathimp.org/curriculum/pig.html
Mathematics: Modeling Our World (ARISE)	Unit Web links: http://www.comap.com/highschool/projects/mmow/weblinks.htm

Fig. 9.2. Technology resources for selected Standards-based mathematics curriculum programs

In addition to the technological components shared by curriculum programs, independent websites offer online tools you can use to support implementing Standards-based curriculum programs. For example, the National Library of Virtual Manipulatives (http://nlvm.usu.edu) offers an extensive collection of online tools and instructional materials. The nonprofit organization Shodor, which offers online tools for exploring concepts in mathematics and science through its Interactivate program (http://www.shodor.org/interactivate/), explicitly identifies how teachers can use the online tools to support teaching particular lessons from five middle grades curriculum programs (http://www.shodor.org/interactivate/textbooks/).

References

Common Core State Standards Initiative (CCSSI). *Common Core State Standards for Mathematics.* Washington, D.C.: National Governors Association Center for Best Practices and the Council of Chief State School Officers, 2010. http://www.corestandards.org.

Mathematical Sciences Education Board. *Everybody Counts: A Report to the Nation on the Future of Mathematics Education.* Washington, D.C.: National Academies Press, 1989.

National Council of Teachers of Mathematics (NCTM). *An Agenda for Action: Recommendations for School Mathematics of the 1980s.* Reston, Va.: NCTM 1980.

———. *Curriculum and Evaluation Standards for School Mathematics.* Reston, Va.: NCTM, 1989.

———. *Principles and Standards for School Mathematics.* Reston, Va.: NCTM, 2000.

Author Biographies

Fran Arbaugh is a former high school mathematics teacher. She received her doctorate in mathematics education from Indiana University, Bloomington, and was on the faculty at the University of Missouri from 2001 to 2009. She currently is a faculty member at The Pennsylvania State University, where she works with preservice secondary school teachers as well as mathematics education master's and doctoral students. Her research interests in mathematics teacher education focus on designing and implementing teacher development programs and include how and what teachers learn from these programs.

Corey Drake is an associate professor of elementary mathematics education at Iowa State University. In research and teacher education, her primary interest is supporting teachers in learning to incorporate new resources into their teaching. These resources include family and community funds of knowledge as well as new curriculum materials, policies, and teaching practices. She is the principal investigator for a National Science Foundation–funded CAREER project studying teachers' uses of curriculum materials as those uses apply to particular developmental and policy contexts. Her recent publications span mathematics education, curriculum studies, and teacher education.

Susan M. Hagen is a senior instructor in the mathematics department at Virginia Tech, where she advises and teaches students in the secondary mathematics education program. She has led many funded projects focused on mathematics teachers' professional development. For the past four years, she has been the mathematics resource person for a team-taught, interdisciplinary course on earth sustainability at Virginia Tech. Her current interests center on preservice teachers' proficiency with technology.

Beth A. Herbel-Eisenmann, a former junior high mathematics teacher, is associate professor of mathematics education at Michigan State University, where she is the elementary mathematics subject area leader for the teacher preparation program. Her research focuses on examining written, enacted, and hidden curriculum by drawing on ideas from linguistics and discourse literatures. She coedited the research volume *Mathematics Teachers at Work: Connecting Curriculum Materials and Classroom Instruction*, published by Routledge in 2009. As principal investigator of a National Science Foundation–funded CAREER project, she spent five years collaborating with eight secondary school mathematics teachers who used action research to align their discourse practices better with their professed beliefs. This group produced an edited volume, *Promoting Purpose-*

ful Discourse: Teacher Research in Mathematics Classrooms, which the National Council of Teachers of Mathematics published in 2009. She serves on the editorial board of the *Journal for Research in Mathematics Education* and received the Early Career Award from the Association of Mathematics Teacher Educators in 2010.

Tonia J. Land is a former elementary school mathematics and science teacher. Currently, she is a doctoral student in mathematics education at Iowa State University, where she works as a research assistant on a National Science Foundation–funded CAREER project and teaches elementary mathematics methods courses. Her research interests include preparing preservice elementary school teachers, teachers' use of curriculum materials, and teachers' use and development of their pedagogical design capacity for teaching elementary school mathematics.

Gwendolyn M. Lloyd, a professor of mathematics education at The Pennsylvania State University, conducts research about teachers' learning, particularly as it relates to teachers' interactions with mathematics curriculum materials. She is coeditor of the *Journal of Teacher Education,* and she formerly served as chair of the editorial panel of the *Journal for Research in Mathematics Education*. She coedited the research volume *Mathematics Teachers at Work: Connecting Curriculum Materials and Classroom Instruction* and has cowritten two publications in the National Council of Teachers of Mathematics Essential Understanding series. Until 2009, she was a faculty member in the mathematics department at Virginia Tech, where she received the Alumni Award for Teaching Excellence in recognition of her work with preservice teachers.

Gina J. Mariano earned her Ph.D. in curriculum and instruction with a concentration in educational psychology at Virginia Tech. She currently works as an assistant professor of psychology at Troy University in Alabama. Her primary research interests include students' learning in online environments, individual differences in students' learning in traditional and nontraditional settings, and faculty's development in online teaching improvement.

Andrea V. McCloskey is an assistant professor of mathematics education at The Pennsylvania State University. She formerly taught mathematics in middle and high schools and earned her Ph.D. from Indiana University, Bloomington. Her research interests include designing innovative ways to prepare and support elementary school teachers in teaching mathematics and developing ways to assess teachers' knowledge for teaching.

Author Biographies

Vanessa R. Pitts Bannister is an assistant professor of mathematics education and mathematics at the University of South Florida Polytechnic. She was formerly a faculty member at Virginia Tech, where she taught secondary methods courses and a course on diversity and equity issues in mathematics education and was the program leader for the secondary mathematics licensure program. She completed a postdoctoral fellowship with the Diversity in Mathematics Education Project at the University of California, Berkeley. Her research interests include teachers' and students' knowledge in algebra and rational numbers, teachers' pedagogical and content knowledge concerning curriculum materials, and equity and diversity issues in mathematics education.

Elizabeth Winarski is a teacher in a multiage kindergarten and first-grade classroom at the Project School in Bloomington, Indiana. She earned a master's degree in elementary education from Indiana University and continues to be involved in research projects related to helping young children engage in mathematical discussions.